Seven Wonders of the Cosmos

This book conveys the thrill of observing strange and surprising features of the universe and the satisfaction gained by understanding them through modern science.

Using simple analogies and a wealth of illustrations, Professor Narlikar skilfully steers us through a cosmic journey of discovery, starting from the Earth and solar system and stepping out to the farthest reaches of the universe. Each of the seven wonders represents a range of mysterious phenomena or a class of spectacular events or remarkable cosmic objects that have challenged human curiosity and often defied explanation.

The first wonder begins when we leave the Earth. Questions are raised such as: Can we see the Sun rise in the west, or find the sky dark despite the blazing sun? The second wonder concerns the giants and dwarfs of the stellar world, and how stars are born, evolve and die. The third wonder concerns the catastrophic event when a massive star explodes, and how the death of one star can trigger the birth of a new generation. The fourth wonder is pulsars, the ultimate timekeepers of the cosmos, the fifth is about strange effects of the force of gravity, the sixth is about illusions of space, and the last is the majestic expansion of the universe as a whole. Finally, we look at other unsolved cosmic mysteries and speculate on what the eighth wonder may be.

With lucid prose and humorous anecdotes, the author weaves together a host of exciting recent discoveries in astronomy and shows us how these are motivating astronomers to unravel the wonders of tomorrow.

JAYANT VISHNU NARLIKAR was born in Kolhapur in India in 1938. He graduated from Banaras Hindu University in 1957. He then studied mathematics at Cambridge University, graduating with the highest honours and the Tyson Medal for astronomy. He continued in Cambridge as a research student of Fred Hoyle, being awarded a Ph.D. in 1963. In 1976 he was awarded the Sc.D. degree of Cambridge University.

In 1963 Jayant Narlikar became Fellow of King's College, Cambridge, and in 1966 joined Fred Hoyle's newly established Institute of Theoretical Astronomy at Cambridge. He returned to India in 1972 to the Tata Institute of Fundamental Research as Professor of Astrophysics. In 1989 he moved to Pune to set up the Inter-University Centre for Astronomy and Astrophysics.

Jayant Narlikar has established world-wide acclaim for his research in gravitation and cosmology, often siding with the minority view in some of the major debates. He is also well known as a popularizer of science and as a public speaker on scientific topics. In 1996 he was awarded the Kalinga Prize by UNESCO for science popularization. He has several technical and popular books to his credit and also enjoys writing science fiction as a form of relaxation.

D1059234

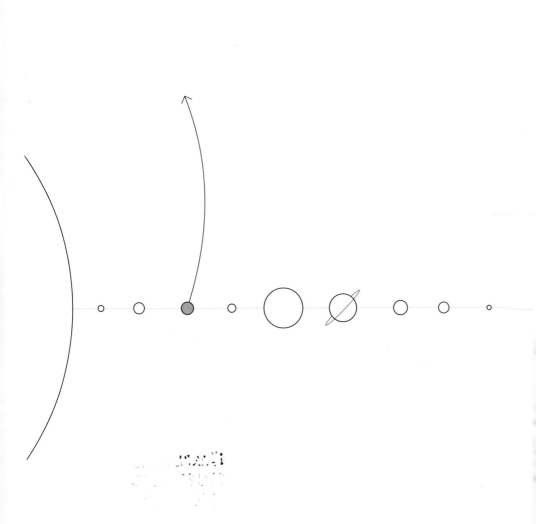

Seven Wonders of the Cosmos

JAYANT V. NARLIKAR

PUBLISHED BY THE PRESS SYNDICATE OF THE UNIVERSITY OF CAMBRIDGE
The Pitt Building, Trumpington Street, Cambridge, United Kingdom

CAMBRIDGE UNIVERSITY PRESS
The Edinburgh Building, Cambridge CB2 2RU, UK   http://www.cup.cam.ac.uk
40 West 20th Street, New York, NY 10011-4211, USA   http://www.cup.org
10 Stamford Road, Oakleigh, Melbourne 3166, Australia

First published 1999

Typeface TrumpMediaeval 9.35/12.5pt   *System* 3b2   [KW]

Printed in the United Kingdom at the University Press, Cambridge

A catalogue record for this book is available from the British Library

Library of Congress Cataloguing in Publication data
Narlikar, Jayant Vishnu, 1983–
Seven Wonders of the Cosmos / Jayant V. Narlikar.
p. cm.
ISBN 0 521 63898 4 (pbk.) – ISBN 0 521 63087 8 (hc.)
1. Astronomy–Popular works.   I. Title.
QB44.2.N37   1999
520–dc21 ' 98-35561 CIP

ISBN 0 521 63087 8 hardback
ISBN 0 521 63898 4 paperback

who may find the real cosmos more wonderful
than the fairy tales I told them.

Contents

Preface

The idea of this book arose out of my public lectures in astronomy. I have always found that a lay audience is highly receptive to information about the universe in the large, provided it is presented in as non-technical a form as possible. In presenting seven wonders of the cosmos to the lay reader I have been conscious of this requirement.

My choice of seven wonders, including the order in which they are presented, perhaps needs some explanation. I have started the cosmic journey from the Earth and the solar system and progressively moved outwards. Each wonder is not a single object, but a subject area.

Thus, the first wonder deals with some unexpected phenomena encountered once one leaves the narrow confines of the Earth. This is followed by the second wonder, the evolution of stars, the most common objects in the sky visible to the naked eye. The third wonder is about exploding stars and the fourth about what is left behind such explosions.

The fifth wonder encompasses the increasingly dominant role of the force of gravity as one encounters more and more massive objects, such as black holes, quasars and active nuclei of galaxies. The sixth wonder tells us about the strange tricks nature may be playing on the astronomer, by producing illusions on a grand scale.

The seventh wonder is the expanding universe and the astronomer's attempts to piece together its history and speculations about its future. Did it start with a big bang? Will it dissolve into nothingness or end in a big crunch? We present some facts and some speculations.

While the epilogue lists some unsolved mysteries, to me the greatest wonder has been the success achieved by the methods of science in coming to grips with cosmic mysteries. Why should scientific laws discovered over three centuries on this tiny planet apply to a history of billions of years of a vast universe? The exciting fact is that they do. And I hope through this book I can share my thrill with the reader.

I wish to thank Adam Black of CUP for encouraging the writing of this book, the three anonymous referees for their constructive suggestions about its style and contents, Santosh Khadilkar, Ram Abhyankar and Prem Kumar for help in preparing the manuscript and illustrations and my wife Mangala for giving a reader's response as well as for helping with the artwork.

I am grateful to Somak Raychaudhury for help in acquiring some of the recent images in this book.

Jayant V. Narlikar
Inter-University Centre for Astronomy and Astrophysics
Pune

Prologue

This book attempts to provide glimpses into the currently exciting areas in astronomy and astrophysics.

The seven 'wonders' described here are not individual objects; they represent a range of mysterious phenomena, a class of spectacular events or a population of remarkable cosmic objects. The attempt to understand them has posed great challenges to human curiosity and ingenuity.

Although a connected theme runs through the seven wonders, each can be read independently.

I hope that, through these wonders, the reader will share the excitement of investigating the cosmos with the professional astronomers who observe and theorize about it.

Leaving terra firma ① 1

THE DAY I SAW THE SUN RISE IN THE WEST

It was a wintry day in 1963, December 14 to be exact, when I saw the Sun rise in the west.

No, I am not trying to pull a fast one! The event actually happened just as I have stated above. But, to restore credibility to the above statement I should elaborate the circumstances. So here is the full story . . ..

It happened while I was on a British Airways flight out of Heathrow, bound for Chicago. I was in a window seat of the Boeing 707, and next to me was the astronomer David Dewhirst of the observatories of Cambridge University. We were both heading for Dallas, Texas for an international symposium on gravitational collapse and relativistic astrophysics.

The sky was clear, of course, at above thirty thousand feet and I had been looking out of the window at the crimson hues towards the southwestern horizon and had seen the Sun go down. A post-lunch stupor was setting in and I was about to doze off for a nap, when David Dewhirst suddenly spoke out. 'Look, there is the Sun coming up again: I am sure I saw it go down below the horizon a few minutes ago.' Even his normally matter of fact way of speaking betrayed suppressed excitement.

I looked out of the window. Sure enough the Sun had indeed come up on the southwestern horizon. And as we both watched over the next few minutes it even rose perceptibly. But this unique spectacle did not last long: the Sun halted in its tracks and finally went down as the plane changed direction southward. It was quite dark when we began our descent into the O'Hare Airport area.

This was the spectacle David Dewhirst and I witnessed that evening. It was an experience I shall never forget.

Why did the Sun rise in the west?

The answer to the question does not call for miracles or for optical illusions. The sight we witnessed was of a real and perfectly natural event that has a perfectly reasonable explanation. And this example demonstrates how different our experiences can be once we leave Mother Earth.

First, let us try to understand why, every day, we see the Sun rise in the east and set in the west. Or, for that matter, why we see the stars move across the night sky from east to west. Today a primary school pupil knows the reason: the Earth spins about its north–south axis and viewed from this moving platform the starry heavens appear to rotate in the reverse direction. This is just like the way a rider on a merry-go-round perceives the surrounding trees and houses go round, in the reverse direction. In order that we see the Sun and the stars go east to west, the Earth itself must be a gigantic merry-go-round spinning from west to east.

Simple! With the help of a globe anyone can understand this hypothesis; but it took mankind millennia to accept it as the true explanation. Let us digress a little and take a peep into recorded history.

'Eppur si muove'

More than two thousand years ago the Greeks, who then had the most sophisticated civilization in Europe, believed that the Earth is immovable and it is the cosmos that revolves round it. Imagine the sky to be a sphere with the stars stuck on it, with the Earth at the centre of the sphere. The Sun and the planets also were supposed to revolve around the Earth though at distances closer than the stars.

Based on a superficial examination of our own experience this belief seems entirely reasonable. Figure 1.1 shows the circular trajectories of stars photographed by a camera that was kept exposed throughout the night. Notice that if a typical star is viewed at any given time it appears like a point source of light. Its position changes slowly, and this is hardly perceptible if we stand and watch it only for a few minutes. However, if we look after a couple of hours, it, along with other stars, will have shifted. The camera in Figure 1.1 has captured the continuous change in the position of each star so that we see a circular track instead of a point source. Compare this figure with, for example, Figure 1.2, which captures the headlights of moving cars in a busy city. In daytime

Figure 1.1: Circular paths of stars in the southern hemisphere, photographed against the backdrop of the Anglo-Australian Telescope. Had there been a pole star in the south, it would have appeared as a point at the centre of these stellar arcs (photograph by David Malin; copyright, Anglo-Australian Observatory).

we likewise see the Sun moving along a circular trajectory from east to west, only it is too bright to be caught on a camera! Thus, to an observer on the Earth, it was perfectly natural to assume that the Earth is fixed and the whole cosmos is revolving.

Figure 1.2: Headlights of cars show straight tracks in a busy thoroughfare. (Compare with the stellar tracks in Figure 1.1.)

But there was one thinker who argued differently. The Greek intellectual, Aristarchus of Samos (*ca* 310–230 BC) argued that these observations can be understood more simply by assuming that it is the Earth that spins from west to east and that the cosmos is in fact non-rotating. Aristarchus, whose work was lost with the destruction of the famous Alexandrian Library also believed that it is the Earth that goes round the Sun instead of the other way round (see Figure 1.3). But his ideas found few takers and for good reasons too. Let us see why.

First, take the example of the merry-go-round. A person standing on it experiences an outward force that tends to push him away from the centre. It is the same effect that we feel when riding in a car that rounds a bend at fast speed . . . we are thrown away from the centre of the bend. So, if we are standing on the spinning Earth why are we not thrown out away from the axis of spin? This question could not be answered in Aristarchus's time.

Second, consider the following simple experiment in a park. View a tree from a distance of say, 50 metres. Now walk about 10 metres side-

Figure 1.3: Aristarchus
of Samos (photograph
by courtesy of Spiros
Cotsakis, of Samos).

ways from the original direction to the tree and look at it again. Its
direction against the background of other trees further back will appear
to have changed. So if we look at a star today and then six months later,
its direction will appear to have changed relative to the background of
other stars farther away *if the Earth has moved during these six
months from its earlier location.* Indeed Aristarchus expected this
result and, to substantiate his hypothesis, he did try to look for it but
could not find it.

So on either count his hypothesis failed. But today we know that he
was right after all, despite these objections. The reason why we are not
thrown away from the spinning Earth is that the magnitude of this force
is very small compared to the pull the Earth exerts on all of us, the pull
of gravity. Because of the force of gravity we are attached to the surface

of Earth and tend to fall back on it if we attempt to jump up and away from it. This is the force that makes us 'feel our weight'. Compared to gravity, the force arising because the spin of the Earth apparently throws us outwards is negligible, being only about three parts in a thousand at the equator and even less at higher latitudes.

As regards the second effect, Aristarchus had grossly underestimated the distance of the stellar bodies and his estimates of the expected changes in the directions of a star were far above the actual changes. (In our analogy of viewing trees from different locations, we know that the direction of a distant tree hardly changes as we change our viewing location whereas that of a nearby tree changes perceptibly.) Thus the direction of a star does change as we view it after six months, but nowhere near as much as Aristarchus expected. And the actual changes in the stellar directions were much too small to be measurable by the purely visual naked-eye techniques available in his times.

The effect that Aristarchus was expecting to see is known today as *parallax* and, with the help of modern telescopes, it has been measured for the relatively nearby stars. Indeed the first measurements of parallax of stars were carried out by the German astronomer Friedrich Wilhelm Bessel in 1838 for the star 61 Cygni, more than two thousand years after Aristarchus! How tiny was the observed change of direction? If we use the familiar degree as the measure of an angle, then the observed change was through about the *ten thousandth part of a degree*! This was far beyond the capability of measurements of ancient Greeks in the days of Aristarchus. No wonder that the contemporaries of Aristarchus found no change of direction for any star as predicted by him. It is not uncommon in the history of science that a scientist with a correct hypothesis which, however, goes against the prevalent belief has had to face derision or neglect if the hypothesis were ahead of his or her times. The irony is that when these ideas finally do get verified and accepted, the identity of their originator may have been lost in the mists of history.

A similar experience awaited the fifth-century Indian astronomer Aryabhata who sought to explain the observation of westward-moving stellar bodies with the analogy of a boat going down a river. The boatman sees the fixed objects on the banks moving backward in analogy to the fixed stars observed from a spinning Earth. Historical records are rather vague, but it seems that Aryabhata was driven by ridicule out of his native state Bihar in North India and had to migrate to the western state of Gujarat from where he had to go out again, to ultimately settle in the southern state of Kerala. Not only that; even his successors in the following centuries sought to push Aryabhata's remarks under the carpet by either discounting their authenticity or 'reinterpreting' them in more conventional terms.

The cultural barriers that existed in Europe and Asia prevented the acceptance of the modern view until the seventeenth century. In the Middle Ages the concept of a fixed earth had acquired the status of religious dogma. The works of Nicolaus Copernicus and Galileo Galilei finally brought about a revolution in thinking but, again, not during their lifetime. Copernicus (1473–1543) argued that not only does the Earth spin about its axis but it also goes round a fixed Sun. His book *De Revolutionibus Orbium Celestium*, which gave a complete description of how all planets, including the Earth, orbit the fixed Sun had a hostile reception as it was widely believed to be against religious tenets.

Galileo (1564–1642) argued even more forcefully for the Copernican theory and was hauled before the Inquisition for propagating heretical views. In the interest of his own survival Galileo recanted but privately continued to believe in the Copernican moving-Earth hypothesis. After the recantation he is believed to have muttered to himself 'Eppur si muove', meaning *but it (the Earth) does move.*

The mystery explained

With this digression over, let us return to the sunrise problem. We will follow Copernicus and Galileo and work with the spinning-Earth model. Figure 1.4(a) shows the latitude circle of Chicago. This runs west to east all round the globe and passes through the location of Chicago. Draw a tangent to the circle. As the globe revolves, this tangent changes direction in space. In Figure 1.4(a) it has the Sun below it, that is, the Sun is below the eastern horizon and therefore not visible. A little later, as shown in Figure 1.4(b), the tangent line touches the Sun and so this represents the sunrise, while in Figure 1.4(c) the Sun is above the line, that is, above the horizon. Thus, the rising of the Sun over in the East is entirely understood by the spin of the earth from west to east. Similarly we can explain the setting of the Sun as the upward movement of the horizon from below.

Now imagine that the spin is reversed! That is, instead of west to east the Earth spins from east to west. Then, by an exactly similar argument, we can deduce that on an Earth spinning in this new fashion the Sun would rise in the west and set in the east.

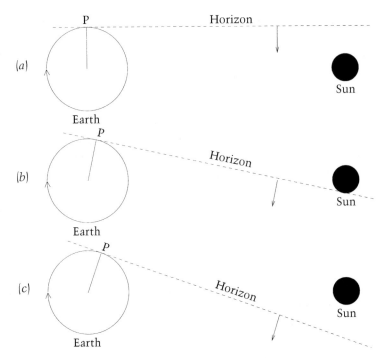

Figure 1.4: Viewed along the south–north axis of the Earth, the latitude circle rotates in a clockwise fashion. In (a) we see a tangent drawn eastward which represents the horizon with the Sun below it. In (b) the Sun is just above the horizon at the time of sunrise while a little later in (c) the horizon has moved further so that the Sun is above it.

But there is a snag in the reasoning developed so far. We cannot in reality reverse the spin of the Earth. So what is the use of this imaginary argument? How can it explain a real experience like the one David Dewhirst and I had? It can, provided we add to it the one clue as yet not used. The clue is: *we were travelling in a jet plane from east to west.* What if our speed in the westerly direction exceeded the Earth's speed in the easterly direction?

An analogy will help. When you stand on a moving belt in an airport you are carried along in the direction of the belt's motion without having to walk on it. If you are in a special hurry you walk on the belt in the same direction and increase your effective speed. But suppose, just to be perverse, you decide to walk in the opposite direction. Then, unless you walk (or run) fast enough you are still moving in the direction of the belt. If you do run sufficiently fast, however, you can effectively reverse the direction of motion.

Substitute, in this argument, the spinning Earth for the moving belt and a jet plane for running, and you will get the idea. If your jet plane can fly faster than the west-to-east motion of the Earth, you will simulate the effect of an Earth revolving in the opposite direction. But just how fast does your jet have to fly to achieve this effect?

Suppose you were flying above the equator. Geographers tell us that the equatorial circumference of the Earth is approximately 40 000 km. In one day the Earth spins about its axis once and so a point fixed on the Equator moves 40 000 km in 24 hours. Simple arithmetic tells us that this corresponds to an average speed of about 1667 kph. A supersonic plane like Concorde can exceed this limit but not a typical 707 or a jumbo jet. Commercial jet planes reach a speed just below 1000 kph. Thus it is not possible for a jumbo jet to match or exceed the Earth's rotational speed while flying over the Equator.

But, the situation gets easier at higher latitudes. Our plane while flying from London to Chicago followed a route that took it over the southern tip of Greenland. The route takes the plane to latitudes higher than those of London and Chicago. Thus, when it passed over Greenland, it must have touched or even exceeded the latitude of 60 degrees. As shown in Figure 1.5, at this latitude, B, the circumference of the Earth is about 20 000 km and hence the west-to-east speed of a fixed point on this circle is less than 850 kph, which can be easily surpassed by a jet flying east to west.

This is what happened to my plane that December evening, which is why I was able to see the Sun rise in the west.

DARKNESS AT NOON

The photograph in Figure 1.6 shows the Sun shining in a dark sky. Yes, the shining ball of light in the photograph is the Sun. But what has happened to its light that bathes the entire sky in blue? Has the Sun lost its power to illuminate the environment?

We know of one circumstance, in itself a rare occasion, when the Sun is in the sky and yet it is dark: a total solar eclipse. But then the solar disc is covered by the Moon, which blocks the sunlight; thus we do not see the Sun shining as in this photograph. How then do we explain the photograph, assuming that it is authentic?

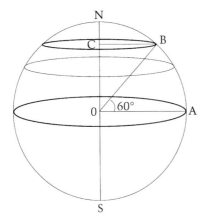

Figure 1.5: The latitude circles get smaller as we move from the equator to the poles. At a latitude of 60 degrees, the circumference of the Earth is half that at the equator. In the figure the length $CB = \frac{1}{2} OA$.

Figure 1.6: The Sun shining in a dark sky (photograph courtesy of NASA).

LEAVING TERRA FIRMA

Before I answer this question, and share with you the secret of how this photograph was taken, let us examine why on a clear day we see the Sun shining in a blue sky. And why, at sunset, does the same sky acquire a reddish tint near the horizon? Even the solar disc is bathed in red at the time of sunset. Why?

Why is the sky blue?

The answer to this question lies in a property of light called *scattering*.

When a ray of light falls on a tiny speck of dust, two things can happen to it. It may be absorbed by the dust particle. Or it may change its direction like a ball bouncing off a piece of stone on the ground. This latter event is called the scattering of light. So when light rays travel through a dusty medium they get partially absorbed and partially scattered as they encounter one dust particle after another. Scattering, however, produces another effect besides changing the direction of a ray.

This effect is called *dispersion* and it simply means that the light gets split into its component colours. We encounter this effect in another context when we pass sunlight through a glass prism (see Figure 1.7). When passing through the prism the light ray changes its direction, first while entering the glass and then while leaving it. Unlike scattering, which turns the light ray at random, this change in direction, called *refraction*, is well determined. It depends on the medium through which the light was travelling earlier (air), the medium it enters

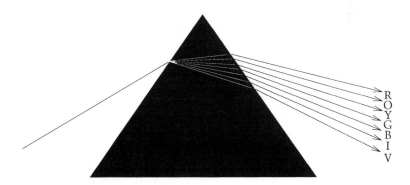

Figure 1.7: The break-up of sunlight into seven colours after it has passed through a glass prism.

(glass) and the *colour* of the ray. It is because of the last property that sunlight passing through the prism gets split into seven colours.

The property of colour can be linked to a basic property of light called its *wavelength*. Thus we can say that amongst all the above seven colours, red light has the longest wavelength and violet the shortest. *What is wavelength?* We will come back to this basic property later in this chapter. For the time being let us concentrate on the readily observable property of colour.

We know that the different colours that make up the sunlight are principally violet, indigo, blue, green, yellow, orange and red (remembered by the acronym *vibgyor*). Of these, violet light bends the most and red the least and the other colours lie in between. The bending angles for different colours can be calculated mathematically. This calculation enables us to understand why the ray of sunlight passing through a glass prism comes out split into a band of seven colours.

In nature the same development takes place when sunlight passes through rain drops, thus giving rise to a spectacular rainbow. As well as suffering refraction, the light rays get internally reflected from the outer boundary of each rain drop, as seen in Figure 1.8. The circular shape of the rainbow comes from the fact that light of each different colour enters our eye from a direction that makes the same angle with the Sun's direction. Thus we see a particular colour distributed in a cricular arc around this direction. Since different colours get refracted by different amounts, we see arcs of different shades of colours, violet being the innermost and red the outermost.

A small fraction of the light rays get internally reflected twice. These rays on emerging from the drop show a second fainter rainbow with an inverted sequence of colours (because of the second reflection).

Scattering by dust in the atmosphere produces much the same effect on sunlight, splitting it into different colours, violet being scattered the most and red the least. The only difference between passing through rain drops and being scattered by dust is that in the latter case the change of direction of the light ray is random and so we do not see a coherent shape like a rainbow. Instead we see the most-scattered colours, those from the violet–indigo–blue family, spread out across the sky whereas the other (less scattered) colours are not spread out that much. And of those most scattered colours blue is the dominating one.

Taking a slight detour from our discussion, we may comment on why traffic lights are red for the stop signal or why, in general, red lights are used to warn against road hazards. For safety of driving, it is neces-

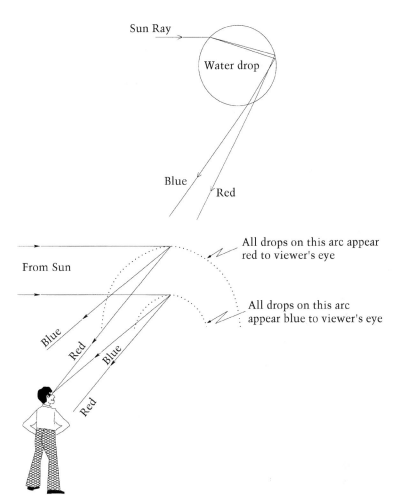

Figure 1.8: A light ray from the Sun enters a raindrop and splits into different colours, all of which get reflected at the far side of the drop and emerge from it in different directions. For an observer, the different colour rays subtend concentric circular arcs, going inwards from red to blue and violet.

sary to ensure that the hazard warning should be easily seen from a large distance, so that fast-moving vehicles can take appropriate braking action. Because red light is scattered the least, it travels farthest in its original direction. Thus in a dusty environment, the stop signal is the one most easily seen from afar *because it is red*.

We can now answer the other question about the redness of the setting Sun. When the Sun is close to the horizon, its light travels through a larger portion of the atmosphere than when it is high up above the horizon. Figure 1.9 explains how this comes about. Therefore the sunlight gets maximally scattered on its way to us. Further, near the horizon its rays graze the usually dusty terrestrial surface before reaching our eyes. In this journey the colour that is scattered the least is the red colour which makes it to us all the way and gives the Sun a reddish appearance.

Can the Sun shine in a dark sky?

Imagine now the opposite situation where the sunlight encounters no dust at all. It would then not be scattered but would travel straight to us. In that case we would see the shining solar disc only and nothing else would be lit up . . . because there is nothing on which the sunlight would fall and be scattered. So if the sunlight happens to travel through an entirely dust-free environment, it will do so without scattering, as in Figure 1.6.

But we live surrounded by a dusty shell of atmosphere and so clearly any sunlight reaching us indirectly has to be scattered. How on Earth, you may ask, is it possible to have the situation I just outlined? The answer is 'Not on Earth!' One has to leave the Earth and go up above the atmospheric shell to attain the above condition. It is here that we may find a truly dust-free resort!

I may now reveal that the photograph was taken in 1993 by an astronaut travelling in the space shuttle Endeavour, high above the Earth's atmosphere. The merit of this vantage point for an astronomer is

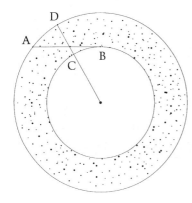

Figure 1.9: Near the horizon the sunlight travels across a larger layer of dust in the atmosphere than when it is overhead. In the figure, the path AB is longer than path CD. (The dusty part is shown mottled.)

obvious. An observer using a space telescope can look at stars or galaxies in the dark sky even in the presence of the Sun! The Earthbound astronomer, however, has to wait until the Sun has set before commencing observations, which must be completed well before sunrise.

There are other advantages of a space telescope by virtue of being above the atmosphere. They are best discussed in the light of our next adventure.

THE STRANGE SIGHTS FROM THE MOON

To see the Sun rise in the west we had to climb aboard a jet plane. To see the Sun shining in a dark sky we needed to go above the Earth's atmosphere. For our next adventure we will go farther and *land on the Moon*. How would the sky look when seen from the Moon? Will we be able to view the Earth from there just as the Moon is observed from here on the Earth?

The picture in Figure 1.10 provides the answer. This is a photograph of the Earth as it appears on the Moon, a photograph taken by the

Figure 1.10: The photograph of the Earth as seen from the Moon taken by the Apollo 11 astronauts (courtesy NASA).

astronauts on the Apollo 11 mission. The crescent of the Earth looks rather like the Moon as seen from here, except that it is comparatively bigger and clearer. *Bigger* because the Earth's diameter is nearly four times that of the Moon and so, seen from the same distance, the Earth would look four times bigger than the Moon. *Clearer*, because the Moon has no atmosphere.

The atmosphere around the Earth plays a doubly inhibiting role so far as astronomers are concerned. First, it absorbs and scatters any cosmic radiation heading towards the Earth, at least partially, and second, by its own movement of air it makes the image of any heavenly source of light unsteady or blurred.

It is because of the lack of atmosphere on the Moon that there is no scattering of sunlight and the Moon's sky is dark despite the shining Sun in the sky, much as in Figure 1.6. Thus the parts of the Moon facing the Sun do light up as the sunlight falls on them; but under a sky that is dark! Figure 1.10 gives us some indication of this very unusual circumstance. Unusual, that is, when judged by terrestrial standards. Because of the highly attenuated atmosphere, sound can hardly travel through it: so, hearing someone talk on the Moon would not be possible.

Let me mention in passing, however, that there was another remarkable aspect of this spectacle of the Earth seen from the Moon that the Apollo astronauts observed: *during their stay, the Earth neither rose further nor set on the Moon: it stayed put where it was in the sky!* We will come back to this strange phenomenon which has a perfectly logical explanation. And, another unusual feature will be that the stars seen here do not twinkle. To appreciate and understand these features we need to probe a little further into what light really is, which is what we shall do next.

Light as a wave

Light has many manifestations. The most familiar form is the sunlight which our eyes use for seeing. This light, as we saw earlier, is made up of seven colours. How would we describe different coloured light to a colour-blind person? Apart from colour, what property distinguishes, say, red light from blue?

In technical jargon we say that it is the *wavelength* that makes the difference: it is longer for red light than for blue. The word wavelength here relates to the fact that light has the form of a wave. What exactly do we mean by a wave? Figure 1.11 shows a

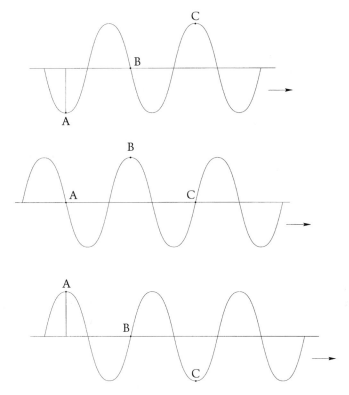

Figure 1.11: The points at *A, B, C, . . .* move up and down *at their locations* without moving to the right; yet, as we compare the distribution curve at successive stages, the overall impression is of a waveform moving to the right. At any given time, the distance between two successive points of maximum upward displacement is called the *wavelength* of the wave. Likewise, the number of up and down movements at any point, per unit time, is called the *frequency* of the wave. The three figures show half a complete cycle, during which the wave advances by half a wavelength.

typical waveform, such as we see when we throw a pebble onto the calm surface of water in a pool. Ripples on the water surface caused by the pebble are seen to move outwards as waves. A closer examination of the water surface shows, however, that the surface moves up and down in such a way that the water particles also are simply moving up and down, but the disturbance as a whole appears to move outwards.

This is characteristic of transverse wave motion. As a wave moves across a medium it causes periodic up and down motions in it. Each such movement up–down–up at any point is called a *cycle*. Let us fix a unit of time, say *one second*, and count these up–down–up motions at a given point. The number of times per second that such cycles occur is called the *frequency* of the wave.

Another characteristic of simple wave motion is that at any time these ups and downs in the medium are evenly spaced, as seen in Figure 1.11. The distance between any two successive *ups* (or *downs*) is called the *wavelength* of the wave. It is this spacing of up and down points and the way they undulate with time that gives us the impression of a moving wave. Figure 1.11 illustrates this principle. A similar on–off switching of neon lights on an advertisement billboard gives the impression of moving letters.

What is it that undulates up and down when a light wave moves across space? For a long time scientists believed that a mechanical medium is necessary to propagate waves. For example, waves in water cause up and down movements in water, sound waves arise from vibrations in air, elastic waves propagate through vibrations in solids and so on. What is it, then, that undulates when light propagates across space? It was assumed that light is a wave moving across an invisible medium called the *aether*. Repeated attempts to detect this mysterious medium, however, failed. The correct answer in fact emerged from the work of James Clerk Maxwell in the 1860s, namely that a light wave is nothing but the transmission of undulating electric and magnetic disturbances across space: an *electromagnetic* wave. The ups and downs in this case are in the intensities of these disturbances across space and time (see Figure 1.12). And the wavelength of light is simply the distance between successive peaks of the electric (or magentic) intensity in space.

Light waves of different colours belong to different wavelength ranges, all of which are extremely short by our daily standard, the metre. A convenient unit for measurement is the *nanometre*, obtained by dividing the metre into a billion (that is, a thousand million) parts. Red light has the longest wavelength, broadly lying in the range 620–770 nanometres (nm), while violet and blue light have wavelengths in the range 390–450 nm. The wavelengths of other colours lie in between.

But what lies beyond this range? Surely nature does not limit itself to the range 390–770 nm? Indeed the limitation to this range is not imposed by nature but by human physiology. In fact, in nature there are several other forms of elecromagnetic waves, which are not detected by the human eye. For example, the waves just on the longer side of red

Figure 1.12: An electromagnetic wave. The undulating electric and magnetic disturbances are each shown by an array of parallel lines. The electric and magnetic disturbances are perpendicular to each other and also perpendicular to the direction of propagation of the wave.

are called *infrared* while those on the shorter side of violet are called *ultraviolet*. Figure 1.13 shows these different types of electromagnetic waves, ranging from radio waves, which have the longest wavelength, to gamma rays, which have the shortest. When we turn on our transistor set and listen to a radio programme, radio waves bring it to us from the radio station.

What is the relationship between frequency and wavelength? It can be seen from Figure 1.11 that during the cycle the waveform advances by one wavelength. The frequency tells us the number of cycles that occur each second. In one second, therefore, the waveform will advance a distance obtained by multiplying the frequency by the wavelength. Since the distance travelled by a light wave in one second is nothing other than the speed of light, we discover that multiplication of the frequency by the wavelength gives us the speed of light. Maxwell demonstrated that the *speed of light through empty space is the same for light of all wavelengths*. This value is approximately 300 000 kilometres per second.

Thus if we know the wavelength of light we can calculate its frequency by the very simple rule we just obtained. For example, a wavelength of 500 nm (for a green light) would have a frequency of about 600 million million! That means that whenever a green light wave passes through empty space, the associated tiny electric and magnetic disturbances execute their up–down–up oscillations 600 million million times every second. (During one such oscillation the wave advances by 500 nm. Therefore in one second, that is, during 600 million million oscillations it will advance a distance of 500 nm × 600 million million = 300 000 km.)

Returning now to the Earth's atmosphere, and how it treats these different forms of light, we may mention that it blocks most other wavelengths apart from visible light, radio waves and some narrow bands in the infrared (see Figure 1.13). To observe the cosmos with telescopes capable of receiving these other wavelength ranges, one needs to go high up into or above the atmosphere. Such telescopes are launched in balloons, rockets or satellites. As we shall see later in this book these telescopes help us capture many other wonders of the cosmos.

Why do stars twinkle?

The second effect of the atmosphere on the stellar images that we referred to earlier was of unsteadiness in the image. Figure 1.7 showed

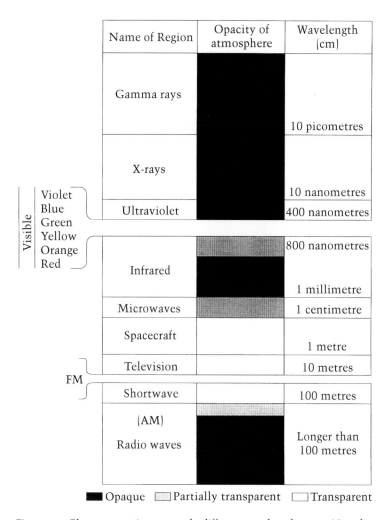

Name of Region	Opacity of atmosphere	Wavelength (cm)
Gamma rays		10 picometres
X-rays		10 nanometres
Ultraviolet		400 nanometres
Infrared		800 nanometres / 1 millimetre
Microwaves		1 centimetre
Spacecraft		1 metre
Television		10 metres
Shortwave		100 metres
(AM) Radio waves		Longer than 100 metres

Visible: Violet, Blue, Green, Yellow, Orange, Red

FM

■ Opaque ▨ Partially transparent ☐ Transparent

Figure 1.13: Electromagnetic waves at the different wavelength ranges. Note that visible light, to which our eyes are sensitive, lies in the middle range, the longest wavelengths corresponding to radio waves and the shortest to gamma rays. The chart also shows the extent to which these waves coming from outer space are absorbed by the atmosphere.

how a ray of light bends as it enters from one medium (air) into another (glass). This effect of refraction, in a more subtle form, operates in the atmosphere for light entering it from above. As the density of the atmosphere increases downward, the light is essentially passing through a changing medium, which gives rise to gradual refraction. Thus the ray changes its direction ever so minutely, as the refraction is small. The effect of this on the stellar image is to shift it slightly.

Now imagine air currents in the atmosphere changing its density distribution slightly, making it shake. The effect, again very subtle, is to make the stellar image shake. So instead of a steady star we see a *twinkling* star. The twinkling, while making the star more spectacular to the eyes of the poet, makes it a more difficult object of study for the astronomer.

To get round this difficulty one direct method is, of course, to get above the shaking atmosphere and set our telescopes there. This is the advantage that the Hubble Space Telescope enjoys over its ground-based counterparts. This telescope not only avoids the dimming of images by atmospheric dust but also avoids their blurring. And of course, on the Moon, too, there being no atmosphere, stellar images will appear bright and clear (see Figure 1.14).

In recent years, however, thanks to advances in technology, ground-based telescopes have begun to employ so-called *adaptive* optics. In this technique, changes in the atmosphere are tracked and the telescope mirror wobbled to compensate for them. By such corrective measures considerable improvement can be achieved in the steadiness of the image.

Why does the Earth appear stationary from the Moon?

After this long detour let us get back to the spectacle of the Earth as seen from the Moon (Figure 1.10). We now understand why it looks so clear that we can even make out some surface features, especially the blue-ness of the oceans. However, if we stay and observe it for a few hours, it will not change its position in the sky. This is strange behaviour, since here on the Earth we are used to seeing the Moon moving across the sky from east to west.

The reason for this peculiar phenomenon is not difficult to under-stand, however. We must take note of an important aspect of the Moon's motion round the Earth. As it moves in a circular orbit, it also spins on its axis in such a way as always to present the same face to the Earth. This is why the other side of the Moon remained hidden to us on the Earth until space technology enabled spacecraft to be sent to the other side. Figure 1.15 shows a photograph taken from a spacecraft sent round to the other side of the Moon.

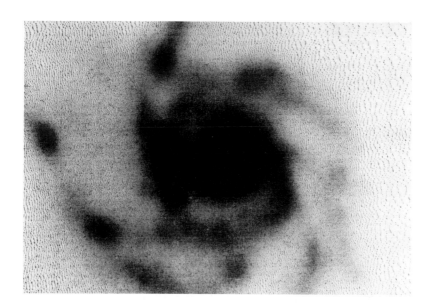

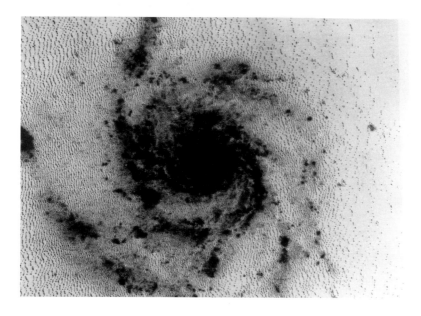

Figure 1.14: The blurring of the image due to the atmospheric turbulence is shown in this simulation in an exaggerated form. In the upper figure we see the image of a spiral galaxy obtained with the 5 m Hale Telescope and in the lower figure the image obtained with the Hubble Space Telescope (HST).

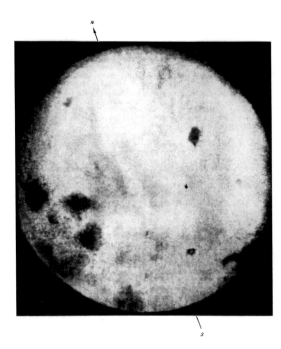

Figure 1.15: The far side of the Moon, photographed in 1959 by the Russian spacecraft Luna 3 for the first time. The Moon spins about its axis in such a way as to present to the Earth the same face while orbiting round it.

This behaviour of the Moon is similar to that of an athlete running in a circle around a flagpost. If the athlete is running clockwise his or her right arm will always be nearer to the flagpost. In order to achieve this the runner is all the time turning round a vertical axis, completing one full turn after running round the circle once. So the runner always finds the flagpost in the same direction, viz. to his (or her) right.

The Moon does the same and so, *if the Earth is visible from the Moon*, it will always be seen in the same direction. The 'if' is important! For, if we happen to be on the 'other side' of the Moon facing away from the Earth, *we would not see the Earth at all.*

REMARKABLE SIGHTS IN THE SOLAR SYSTEM

The Earth when seen from the Moon looks about four times larger than the Moon when seen from the Earth. This is but an example, a relatively modest one, of the wide variety of possible sights in this entire solar system of nine planets and all their satellites. Sights more dramatic than we are used to on the Earth may be in store should we be able to capture them.

In a lecture entitled 'The astronomer's luck' the astrophysicist William H. McCrea once talked of several accidental factors that have intervened in astronomy. His lecture began with a discussion of the apparent sizes of the Sun and Moon as seen from the Earth. Both the solar and lunar discs appear to be of the same size and may give the impression that they are physically equal. The reality is that the Sun's diameter is about 400 times larger than the Moon's. But because the Sun happens to be located so much further away than the Moon, its large size appears reduced to almost exactly match that of the Moon. How much of this is a coincidence can be appreciated with the help of Figure 1.16, which illustrates the geometrical aspects of this situation.

This figure explains what it is that gives us the subjective impression of how large an object looks. In this picture a round object is seen by two observers A and B, A from nearby, B from far away. It is common experience that the object would look much bigger to A than to B. The reason is that the image formed on the retina of the eye of observer A is much larger that that formed on B's retina. This image is basically determined by the angle subtended by the spherical object at the eye. As seen from Figure 1.16, this angle is much larger for observer A than for observer B. An approximate measure of the angle subtended at an obser-

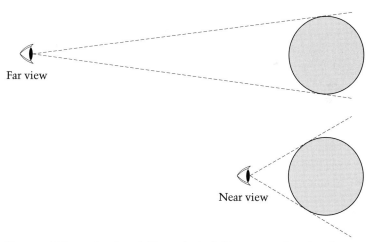

Figure 1.16: The angle subtended by a spherical object at a nearby point of observation is much larger than at a remote point. This angle determines the apparent size of the object to the observer at each of these points, which is why things look bigger when viewed from a closer range. In the text, the far view is that of observer B and the near view that of observer A.

ver by such an object is given by the ratio of the diameter of the object perpendicular to the line of sight to its distance from the observer. In fact this approximation is quite good when the angle is small.

In the case of the Sun and the Moon, we noted that the linear extent of the Sun is some 400 times that of the Moon. Now it so happens that, from where we are, our distance from the Sun is also about 400 times our distance from the Moon. So by the rule just arrived at, the apparent size of the Moon is very close to the apparent size of the Sun. It was this coincidence to which Bill McCrea was referring.

Because the Moon matches the Sun so very well in its apparent size it is possible, on some rare occasions, for it to cover the Sun entirely, thus causing a *total* solar eclipse. These occasions are rare, however. As illustrated in Figure 1.17, Sun, Earth and Moon do not move relative to one another in quite the same plane. The plane of the Earth's orbit round the Sun and the plane of the Moon's orbit round the Earth make a small angle of about 5 degrees to one another. This is why the occasions when the Sun and the Moon are exactly aligned with respect to the Earth are rather infrequent. But before proceeding further into the geometrical aspects of this alignment, consider the following tale from Indian mythology.

In primordial times the gods and demons joined together in a mammoth operation to churn the ocean in the hope of extracting submerged treasures. They had arrived at some prearrangement for sharing what might surface from this exercise and, accordingly, when nectar was found, it was meant for the gods. However, as it was being distributed amongst them, one demon surreptitiously joined the crowd hoping to grab a portion of this elixir. But the Sun god and the Moon god found him out and reported the outrage, whereupon god Vishnu who was distributing the nectar chopped the demon's head off. The demon did not die, however, but survived in two parts, the head being called 'Rahu' and the rest of the body 'Ketu'. Angered with the Sun and the Moon for betraying him, the demon's two remnants decided to eat them. Accordingly, Rahu succeeds in swallowing the Sun and Ketu the Moon on special occasions – but only for brief intervals. These are of course the occasions of solar and lunar eclipses.

This myth took a strong hold of Indian societies and, even today, occasions of eclipses are treated with great caution and awe, largely because to the traditional mind the mythological demons Rahu and Ketu continue to assume reality. It is not uncommon to find the normally busy city streets deserted at the time of a solar eclipse. As seen

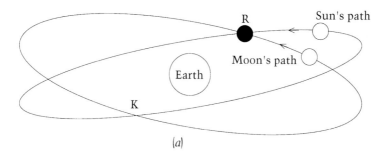

(a)

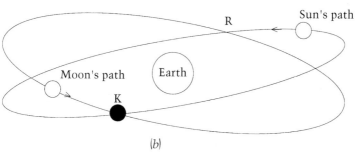

(b)

Figure 1.17: The planes in which the Moon moves round the Earth and the Earth moves round the Sun are inclined at a small angle. Their line of intersection RK defines the directions of nodes. In (a), at a solar eclipse, the Moon is in between the Sun and the Earth and a mythological interpretation is that a demon called Rahu at node R has swallowed the Sun. In (b), at a lunar eclipse, the Earth is in between the Sun and the Moon and the interpretation likewise is that the demon Ketu at node K has swallowed the Moon.

from Figure 1.17, eclipses occur when the Sun and the Moon are at the points on the line where the planes of their orbits intersect. These points are the *nodes* where the mythical demons Rahu and Ketu are supposed to exist!

Because of the relative rarity of total solar eclipses, which occur when and only when the Moon blocks out the Sun, these phenomena have inspired the public mind with such folk tales. On the one hand, if the Moon's apparent size had been much larger than the Sun's, or if the Moon were slightly closer to the Earth, solar eclipses might have been more common and would have lost their rarity value! On the other hand, if the Moon were even a few per cent smaller in size or located a little farther out from the Earth, the total solar eclipse would not have been seen at all. Figures 1.18(*a*) and (*b*) illustrate this argument.

Of all the planets in our solar system and their satellites only the Earth–Moon system enjoys this critical coincidence. Let us pause and reflect what eclipses we would see from the Moon. Would we see the Sun eclipsed by the Earth and the Earth eclipsed by the Moon?

As we noticed before, the Earth seen from the Moon is some four times larger than the Moon seen from the Earth. The Earth therefore subtends a (nearly four times) larger angle at the Moon than the Sun. So if, as seen from the Moon, the Earth gets in the way of the Sun, we would have a total solar eclipse. This happens, of course, when there is a lunar eclipse seen from the Earth! Because on the Moon the apparent size of the Earth is nearly four times larger than that of the Sun, such eclipses

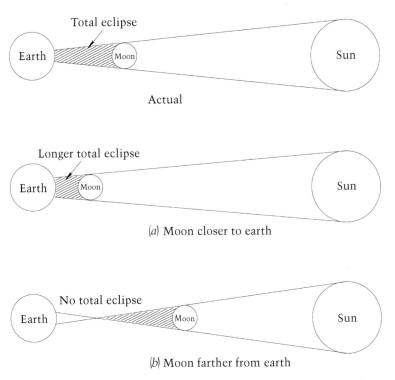

Figure 1.18: (*a*) If the Moon were bigger in size or closer to the Earth, its shadow would cover the Sun more easily, as shown here, thus causing solar eclipses to be much more common events. However, if, as in (*b*), it were smaller in size or farther away from the Earth, it would not be able to block the Sun entirely; thus a total solar eclipse would be rare or impossible.

will be more frequent than solar eclipses on the Earth and they will last much longer. They will not, however, be as impressive as the total solar eclipses on the Earth. Why?

Recall that, because of the scattering of sunlight by the Earth's atmosphere, the sky is bathed in sunlight on a normal day. At the time of the eclipse, the sky becomes dark for a very short duration. This is an impressive spectacle, with stars appearing, the temperature dropping and the corona lighting up around the covered solar disc. On the Moon, however, the sky is always dark, whether the Sun is there or not. Thus the covering of the Sun would not be such a dramatic spectacle on the Moon as it is on the Earth.

What about an Earth eclipse seen from the Moon? A total eclipse will happen when the shadow of the Moon envelops the Earth, that is, when there is a solar eclipse on the Earth. However, the Earth is much too big to fall entirely within the Moon's shadow cone. And so we would see only a very partial eclipse of the Earth.

So, as William McCrea observed, astronomers are indeed lucky that the Sun and the Moon appear in the sky to have very nearly the same size. This coincidence is not found for any of the other planets and their satellites. From some of the other planets, however, other impressive sights may be routinely observed.

View from Io

We will now imagine one dramatic example of this. For this purpose we have to go to the vicinity of the planet Jupiter.

Jupiter is the largest planet in the Solar System, having a diameter about 12 times that of the Earth. It has thus a surface area that is about 150 times that of the Earth and a volume that gobbles up nearly 2000 Earths!

Jupiter has 16 satellites and one of the inner satellites is Io. Io is about the same size as the Moon and moves round Jupiter at a distance that is about ten per cent more than the Earth–Moon distance. Now think of how Jupiter would look if it were seen above the horizon of Io. Like the Earth's seeming stationary to an observer on the Moon, Jupiter would appear fixed above Io. But how large would it look? Physically the radius of Jupiter is more than eleven times that of the Earth and we recall that the Earth is nearly four times the size of the Moon. *Calculations show that Jupiter as seen from Io would be forty times larger than Moon seen from the Earth!*

In Figure 1.19 we see a photograph of Io beside Jupiter, which may convey some idea of the gigantic size of the planet when viewed from its nearby satellite.

Of course, this is what scientists call a *thought experiment*, an experiment that is only imagined but not actually carried out. Indeed many thought experiments are such that they *cannot* be carried out. The Io viewing of Jupiter is one; the satellite is not sufficiently hospitable for us to land there and actually carry out this experiment. But that should not prevent us from *imagining* how Jupiter would look from Io. Indeed, during the course of this account of seven wonders of the cosmos we shall have occasion to carry out more of these thought experiments.

Adieu to the Earth

And there we leave the first of our seven wonders. Having been born and brought up on this Earth we are accustomed to a certain pattern of natural phenomena. We have to appreciate that these phenomena, diverse, impressive and majestic though they are, are inevitably limited

Figure 1.19:
Photograph of Jupiter taken by the Voyager II spacecraft on 10 June 1979. The tiny satellite Io may be seen to the right (courtesy of NASA).

by the size, environment and other physical properties of this planet. The set of 'wonders' described in this opening chapter has shown us glimpses of what lies out there once we leave our terra firma. No doubt with the progress of human endeavours in space, the twenty-first century will add further to these glimpses. But as we move on to the remaining wonders we will find that they lie so far away there there is no way of watching them from close at hand. Instead one has to rely on the remote observing provided by our astronomical techniques. Even though these cosmic events are taking place a very great distance away, we are able to appreciate their majesty and estimate their enormous scale.

Giants and dwarfs of the stellar world $\;$ 2

We discussed the first set of cosmic wonders in Chapter 1, where we encountered highly unfamiliar and unusual situations once we left the confines of Earth. As we mentioned towards the end of that chapter, the rest of the wonders in this book relate to the more and more remote parts of the universe, parts which are far too remote for us even to contemplate a real cosmic voyage by a spaceship containing human beings.

Just to put this point in perspective, let us compare the first ever sortie by Earthmen to another habitat in the solar system. This 'space mark' was achieved by Neil Armstrong and Edwin Aldrin on 20 July 1969, when they stepped onto the Moon. As Armstrong put it at the time, this 'one small step for man' was 'a giant leap for mankind'. Indeed it was a historic moment when someone from the Earth, for the first time ever, stepped onto an extraterrestrial surface.

The journey to the Moon on this Apollo 11 mission took about seventy-three hours each way. How far is the Moon from the Earth? The distance can be given in kilometres or miles: but let us use a different unit, more suited to astronomical distances. The fastest means of signal transmission available in nature is light. Light travels a distance of approximately 300 000 kilometres in one second. So we can estimate the distance of an astronomical object by the *time* taken by light to travel from there to us. A distance of *one light second* is thus the distance which light travels in *one second*. This, as we just saw, is 300 000 kilometres. On this scale, the Moon is about one and a quarter light seconds away.

So here we have a problem in arithmetic, which could very well appear in a school textbook. The problem is as follows.

The nearest star to the Earth, after the Sun, is *Proxima Centauri*, located about four and a quarter light years away. How long will our lunar spaceship take to reach this star?

To paraphrase the problem, note that the Apollo spaceship took seventy-three hours to traverse a distance that light covers in a second and a quarter. Let us suppose that a more modern spaceship will cover that distance in fifty hours today. So, how long will this spaceship take to cover a distance which light takes four and a quarter *years* to cover? Anyone who has not forgotten the standard method of ratio and proportion can solve this problem. The answer will come as something of a shock: *it is around six hundred thousand years*. Obviously, we require a technology for interstellar travel vastly superior to what we possess today!

Nevertheless, even if cannot go there, astronomy allows us to observe and appreciate cosmic wonders located far away. In this chapter we will take a look at the star-studded night sky and see how, with the help of their telescopes and scientific theories, astronomers have succeeded in unravelling the physical nature of stars. And the picture they have unfolded will take the breath away.

What are the means by which it is possible to study and understand stars so far away? We will describe this success story of the science of modern times in small steps. It is one of the wonders of our cosmic voyage.

OF STARS AND HUMANS

Let us imagine the following scenario. A spaceship carrying advanced extraterrestrials is approaching the Earth. The ETs are considerably more advanced than we humans; but they are cautious about landing the ship on this planet. Before landing *en masse*, they want to know more about us: about how human beings are born, how they grow, how they live their life and how they die. To find these details they beam one of their number down to Earth with instructions to find out as much information as possible in a comparatively short time, say, an Earth week.

How will the ET proceed to find out the facts? Clearly, the obvious way, proceeding to the maternity ward of a hospital to see a child being born and then following his or her life for the full lifespan of seven or eight decades, will take too long. Also, at the end of the exercise the ET would have found all about only a single member of the human population. Knowing the variety of even the human species, this single case might well be quite misleading.

Indeed, the practical way open to the ET is to resort to survey and statistics. Going to a big city and inspecting the humans there, the ET will get a feel for the variety of the species. There will be large specimens, small specimens, tall ones, short ones, with skins of different colours and textures, hair of different colour, length and thickness, and so on. Collecting data on a large enough sample, the ET will be able to deduce something about the development of a human being with age.

For example, Figure 2.1 shows a plot of height versus weight for humans in just such a large group. Notice that there is a thinnish tail to the left of the plot, where both the height and weight are small. Then there is a plateau with no significant growth of height but a large variation of weight. From our understanding of human development, we can say that the left end denotes the period of growth from child to adult, while the plateau shows the period of adulthood. The fact that there are

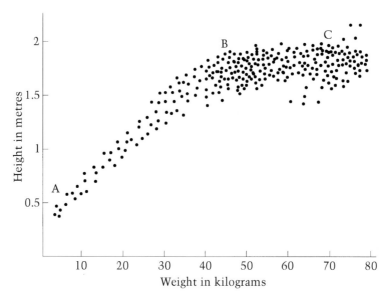

Figure 2.1: This plot shows the weight on the horizontal axis and height on the vertical axis of human inhabitants of a typical town. In Figure 2.4 we show a similar plot for a group of stars. Section AB denotes early growth while BC covers the adult phase.

GIANTS AND DWARFS OF THE STELLAR WORLD

more points on the plateau than on the rising curve indicates that on an average a human being spends a smaller part of his or her life growing to adulthood compared to the lifespan as an adult. Additional data on the texture of skin, quality of hair etc. will give the ET further information about ageing, provided the ET has mastered advanced biology. With these data, therefore, the ET will be able to piece together a broad story of the typical human life and will also have some idea on the extent of variation amongst humans; this latter, because the study encompasses a large sample.

This scenario of discovery about humans holds the clue to the problems about stars that confront astronomers. The questions that they wish to answer are: *How does a star live out its life? How is it born? How does it acquire its shape, colour and size? Do these characteristics change as the star ages? And, the most fundamental question of them all, what makes a star shine?*

How do astronomers go about answering these questions?

Again, there are two ways open to them. In the Sun, they have a star very close by, which they can observe in great detail. If they keep observing the Sun will they not get an answer to their question?

Hardly! For, during the lifespan of a human being the Sun does not appear to change. Nor has it noticeably changed over the lifespan of the entire human race. Indeed, these spans are so short that they hardly count in the development or *evolution* of a star like the Sun. Moreover, suppose that, like humans, stars are not all similar: can we then get to know all about them by observing only the Sun? Again we need the equivalent of the ET's second method, which surveys a large sample and draws statistical conclusions.

The starry sky indeed presents us with a large population of stars. On a clear night, with the naked eye we can see a couple of thousand of them. There are, of course, far more beyond our unaided vision. With the help of telescopes, photography and modern computer techniques, we can record hundreds of thousands. And these studies reveal that stars are usually found in clusters. That is, instead of an isolated star, typically we find a large group of stars moving around one another. There are reasons to believe that the stars in a cluster were born grouped together although not necessarily all at the same time.

How are stars born?

We will come back to this question in the next chapter. For now we concentrate our attention on stars in a cluster and recall the analogy of humans in a large city.

In Figure 2.1 we had a plot of weight against height for humans. Can we think of a similar plot for stars? Yes. Such a plot is available although it does not involve the 'heights' and 'weights' of stars. Rather, it involves two other features of stars that the astronomer can measure despite the stars' being so far away. This plot was independently thought of by Ejnar Hertzsprung (1873–1967) and Henry Norris Russell (1877–1957) (Figures 2.2 and 2.3) and is now known as the Hertzsprung–Russell diagram or, simply, the H–R diagram.

Figure 2.4 shows an H–R diagram for the nearest and brightest stars. On the horizontal axis is plotted the star's surface temperature and on the vertical axis its luminosity, that is, the rate at which the star is radiating energy. How does the astronomer determine these quantities? We will discuss this in the next section. First, though, we will discuss the broad features of the H–R diagram, Figure 2.4.

Notice that a large number of stars, including our Sun, lie on a thickish band extending from the top left-hand corner down to the bottom right-hand corner. Reading the horizontal scale we find that the temperature of the surface decreases to the right. Thus the stars at the bottom right are relatively cool, say around 4000 degrees Celsius, while those at the top left may be more than three times as hot. The

Figure 2.2: Ejnar Hertzsprung (courtesy of Astronomy Department, Yale University)

Figure 2.3:
Henry Norris Russell
(courtesy of Yerkes
Observatory).

Sun, with a surface temperature close to 5500 degrees Celsius, is somewhere midway on this band.

This band is called the *main sequence*. Like the horizontal band in the human diagram of Figure 2.1, the main sequence represents the longest duration of a star's life. Of course, not all stars are on this band. Quite a few are above it, in the top right-hand corner. These are cooler but much more luminous stars and are called *giants*, for reasons which will become clear in due course. Likewise, the stars below the main sequence to the bottom left are called *dwarfs*. These stars are very hot but at the same time very faint.

We will first look at the physical features of the stars before coming to the question how they acquire them. Indeed, the understanding of stars represents a remarkable triumph for science. This success has demonstrated that the laws of science that we study on our rather

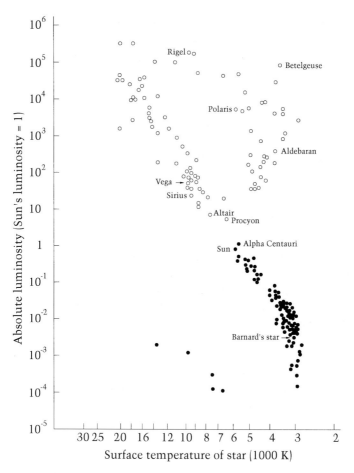

Figure 2.4: The H–R diagram for the nearest and brightest stars. The Sun and a few well-known stars are shown by name. The solid circles correspond to the nearest stars, the open circles to the brightest stars.

small and humble planet are applicable to such large objects as stars located several light years away.

PHYSICAL CHARACTERISTICS OF STARS

Looking at the starry heavens, one's first impression is of an identical set of shining dots distributed all over the sky. A more careful visual examination shows that *they are not all identical*. Some are brighter, some are fainter, some are larger than others, some are bluish in colour while some tend to have a reddish hue. Going beyond naked-eye viewing, the astronomer uses a telescope coupled with some other instrument. The telescope gathers light in large quantities from the source and focusses it at a convenient point. There the instrument takes over. It can use the focussed light to form an image, or analyse it into its spectrum of different colours or measure some of its other properties. We next see how these instruments help in processing the information from stars.

Stellar luminosities

The photographic camera invented in the nineteenth century was a boon to the astronomer, as it enabled faint sources that were otherwise invisible to the eye to be photographed. In a camera, the film can be exposed to the distant light source for a long enough time that sufficient light is collected to form an image. In this way the camera serves as an ideal ally to the telescope in revealing the unseen universe. Not only stars but other fainter cosmic objects have thus become targets for study.

Figures 2.5–2.7 are examples of a few such faint objects in the sky. These are generically named *nebulae*, and the word 'nebulous' in the English language, indicating a vague or hazy object or notion, has come from the name of such images. Notice that unlike stars, which appear as concentrated sources of light, these nebulae do not seem to have a sharp boundary; this suggests that they probably extend beyond what is seen in these pictures. Clearly a faster film and a longer exposure might reveal more.

Modern technology has provided a new device for imaging faint astronomical objects. Known as the CCD (charge coupled device), this instrument has revolutionized astronomical imaging. Shown in Figure 2.8, the CCD keeps track of how the light intensity is distributed on different parts of the imaging surface. It is convenient to use the notion of light intensity being measured in tiny packets called *photons*. This notion is derived from *quantum theory*, which studies the behaviour of matter and radiation at the microscopic level. At this level light, which we have already encountered as a wave, appears to show effects as if it is made up of particles. Photons are thus particles of light, and when they fall on the CCD surface they release electrons from the

Figure 2.5: The North America Nebula (CCD picture by amateur astronomers Dominique Dierick and Dirk De Marche).

surface and these are registered by special counters. Thus more electrons will be released where more photons have fallen, and so counting the electrons gives us an indication of the faint and bright parts of an image. A computer attached to the instrument keeps track of how many electrons came from which part of the surface and then converts the counts into artificial pictures. These pictures use different colours to differentiate between areas of varying counts, much like the contour maps in a geographical atlas.

Figure 2.9 is a black and white version of such a picture. The use of the computer is a great asset to the astronomer studying such images. For, by altering the levels of intensities, the astronomer can highlight certain portions of the image, magnify it, rotate it etc. These operations go under the name *image processing*.

By collecting light from a star, the astronomer can measure what is known as its *apparent brightness*; apparent, because, the image does not contain the full information about the real *luminosity*, that is, the active rate at which the star radiates energy.

An example with light bulbs will illustrate this. Suppose we view an illuminated light bulb of 10-watt power from a distance of 10 metres. We form a certain impression as to its brightness. As we move farther

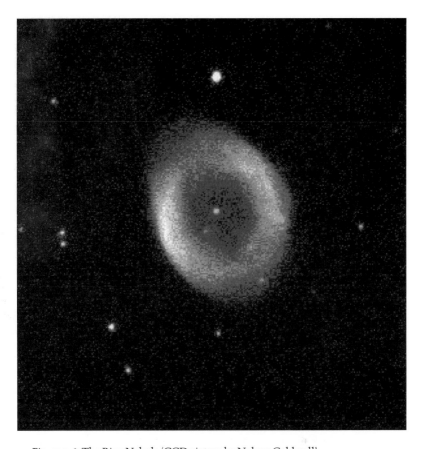

Figure 2.6: The Ring Nebula (CCD picture by Nelson Caldwell).

away from the bulb, it appears to become fainter. From a distance of 100 metres it will appear very faint indeed. We say that the apparent brightness of the bulb decreases as its distance from us increases. At what rate does this decrease occur? To find out, the same experiment may be repeated with a 1000-watt bulb. Although intrinsically it is brighter than the 10-watt one, it too will appear to grow fainter as we move away from it. However, we can ascertain from several trials that its apparent brightness at a distance of 100 metres very closely matches the apparent brightness of the 10-watt bulb viewed from a distance of 10 metres.

This means that to compensate for a decrease in apparent brightness arising from a *tenfold* increase in distance, we need to boost up the

Figure 2.7: The Orion Nebula (picture by amateur astronomer Jason Ware).

luminosity of the bulb a *hundredfold*. The result may be generalized to what is commonly known as the *inverse square law of illumination*: the apparent brightness of a source of light falls off in proportion to the inverse of the square of its distance from the observer. Or, to put it differently, if we have two soures of light, source A being n times farther away than source B, then for them to appear equally bright to the observer source A must be $n^2(= n \times n)$ times as luminous as source B.

There is a simple way of understanding the inverse square law of illumination. In Figure 2.10 we have a light source A emitting radiation equally in all directions. Such a source is called an *isotropic source*. With A as centre draw a sphere S with radius r. Then the area of the surface of S is $4\pi r^2$, where π is often approximated by the fraction 22/7. Taking the approximation for π we conclude that a sphere of radius 7 metres will have a surface area equal to 616 square metres. However, let us concentrate on the sphere of radius r. Imagine an observer O located on this sphere. How much energy from A will come per second per unit area around this observer? This quantity will define the apparent bright-

Figure 2.8: A charge coupled device, mounted on a card with accompanying electronics.

ness of the source. Since all points on the sphere are treated equally in the sharing of A's energy, and the area occupied by them all is $4\pi r^2$, the amount of radiation from A coming across unit area will be equal to the luminosity of A divided by $4\pi r^2$. The share received by O comes down in proportion to r^2, that is, it drops down as the inverse square of r.

What applies to light bulbs applies also to stars. If we observe two stars A and B, and find that A looks much fainter than B, what can we conclude about their distances? If we know that A and B are stars of equal luminosity, then we can say that A is farther away than B. But if we do not have this extra information then, naturally, we cannot make such an assertion. For example, A and B could be at the same distance, with A much less luminous than B. As it turns out, the stars that we can observe with the naked eye are not necessarily the nearest stars. By and large, they are more luminous and distant stars. Some of the really close

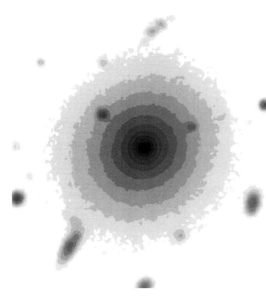

Figure 2.9: A computer-generated picture of the galaxy 0434-225 taken at the Las Campanas Observatory, Chile. The different contours (separating areas of different shades of colour) represent lines of equal intensity (courtesy of Ashish Mahabal).

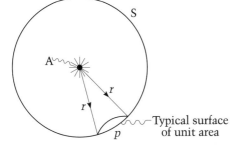

Figure 2.10: The point source of light A emits equally in all directions. A sphere centred at A is shown, across whose surface all light from A is crossing uniformly. Each unit area on the surface has the same amount of light crossing it outwards.

stars are intrinsically so faint (i.e., with such low luminosity) that we cannot see them without telescopic aids.

Normally, with the help of telescopes and such light detectors as the CCD, the astronomer is able to measure the apparent brightness of a source. If measurements of the distance of the source are also possible, then the astronomer can estimate the luminosity of the source. This is done by simply inverting the result obtained just now: multiply the observed apparent luminosity per unit area by $4\pi r^2$, where r is the measured distance to the source.

Let us apply this method to estimate the luminosity of the Sun. The Sun's distance from the Earth is about 150 million kilometres. The amount of light energy coming from the Sun over one square kilometre of area per second is about 1500 megawatts. That is, if we could convert all the solar energy falling on a one square kilometre area, we could use it to run a 1500 megawatt power station. So, by using the above method of calculation we estimate that the Sun has a luminosity of about 400 million million million megawatts! By terrestrial standards it is enormous indeed! But not by astronomical standards, as we shall shortly see.

If we look at the H–R diagram of Figure 2.4 again, we find that the Sun is about halfway towards the top on the luminosity axis. There are stars a hundred times more luminous than the Sun in that diagram. Such stars are to be found both on the main sequence and as giants.

Spectrum of a star

As we mentioned before, we can split the light of a star into seven rainbow colours just as we can split sunlight by passing it through a prism, or through a more sophisticated instrument such as a spectrograph. The different colours correspond to light waves of different wavelengths. In a spectrograph, these wavelengths can be measured.

Let us look at the spectrum of sunlight taken through a spectrograph (Figure 2.11). Over and above the continuum of light, which varies in colour from violet at the shortest wavelength to red at the longest, we notice a series of dark lines. Where do these lines come from?

First discovered in 1814 by Joseph von Fraunhofer (Figure 2.12) and subsequently named after him, the Fraunhofer lines posed a mystery for over a century. It was resolved only when a major revolution took place in the theoretical framework of physics, through the discovery of the quantum theory. Let us try to understand their origin in terms of this theory.

Quantum theory seeks to describe the behaviour of the microscopic structure of matter, on the scale of atoms. A typical atom has a size of the order of a tenth of a nanometre. By the beginning of this century physicists had begun to discover that the Newtonian laws of motion that had been so successful in describing terrestrial as well as astronomical systems did not seem to work in such tiny systems. Let us take as an example the simplest atom, an atom of hydrogen.

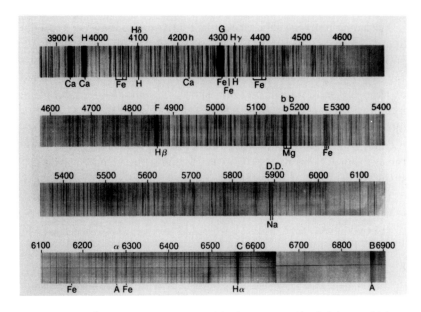

Figure 2.11: The continuous spectrum of the Sun is crossed by dark lines, which were first discovered by J. Fraunhofer. The units are angstroms (Å); $10\text{Å} = 1$ nanometre $= 10^{-9}$ metres.

Figure 2.12:
J. Fraunhofer.

Figure 2.13 shows schematically the semiclassical picture of the hydrogen atom based partly on Newton's laws. It has only two particles of matter, the electron and the proton, both carrying electric charge. The charge on a proton is positive and that on the electron negative; but both charges are of the same magnitude. However, the proton is much more massive than the electron, its mass being some 1836 times the electron mass. The electron is never at rest. It keeps on orbiting the proton, which, being more massive, remains more or less stationary as the electron goes round it. Classical electrodynamics then tells us that such an orbiting electron will lose energy by emitting radiation. And, as it does so, it falls closer and closer towards the proton, the whole process taking time of the order of a million million million millionth part of a second. How then can an atom of hydrogen retain its finite size?

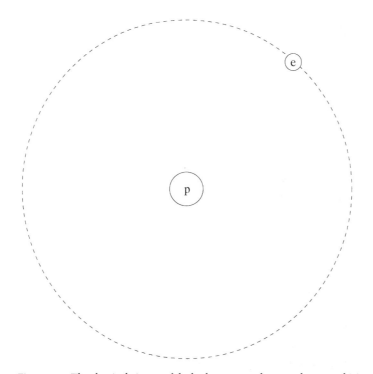

Figure 2.13: The classical picture of the hydrogen atom has one electron orbiting a proton, much like a planet orbits the Sun. If the electron were supplied with more energy its orbit would move outwards in a continuous manner if classical rules prevailed.

It was Neils Bohr, the Danish physicist, who in 1913 offered a resolution of the problem. In the classical case, as the electron lost energy, its orbit shrank continuously, getting eventually to size zero. In Bohr's solution new rules of quantum theory intervened to show that the electron can orbit, without radiating energy, but that the sizes of these orbits form a *discrete set*.

Figure 2.14 illustrates the quantum situation, again schematically. It shows two permissible orbits in which the electron can move. These are successive orbits in a discrete set, the outer orbit having higher energy than the inner orbit. Suppose that the electron is presently in the inner orbit, to enable it to move to the outer orbit, it is necessary to

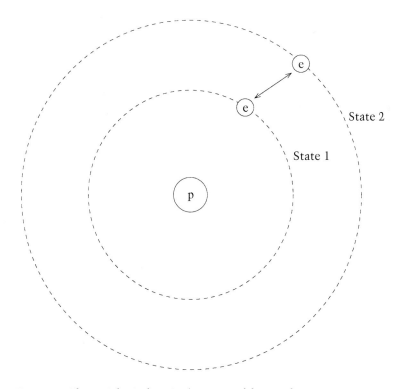

Figure 2.14: The semiclassical version (prototype of the complete quantum version) of the hydrogen atom has the electron moving in one of a specified set of states with different energies. Two such states are shown here. The electrons can jump from one state to another by emitting or absorbing radiation energy. In the figure the energy in state 2 is higher than that of state 1. So an electron in state 1 would need energy from an external source in order to jump to state 2.

supply it with additional energy equal to the difference in energies of the two orbits. *Only if it receives this amount of energy, no more, no less, will the electron move out to the second orbit.*

In practice, this energy may be available to the electron from electromagnetic radiation. Here, quantum theory tells us that the radiation of a specific frequency comes in packets called *quanta*. The rule first stated by the German physicist Max Planck is quite simple. Multiply the frequency of electromagnetic radiation by a physical constant h and you get the quantum of energy. The constant h is a universal constant and is known as *Planck's constant*. It plays a key role in all phenomena described by the quantum theory. Later Einstein introduced the idea of the *photon*, the particle of light, which is the same as the quantum of radiation used by Planck. To see how much energy a photon carries, the following example may be of interest. The photons of a radio wave of 1 metre wavelength carry an energy of about 2×10^{-25} joules (one fifth of a millionth of a millionth of a millionth of a millionth part of a joule!). In red light of wavelength 700 nanometres each photon has an energy around 2.8×10^{-19} joules. Even a gamma ray photon of very high frequency carries a very tiny energy measured by our day-to-day standards. The term 'quantum theory' was in fact derived to emphasize this tiny *quantum* of energy carried by a packet of electromagnetic radiation.

A digression

At this stage, the reader may rightly express bewilderment: isn't electromagnetic radiation made of waves as we stated in Chapter 1? How come it is also described as a collection of particles called photons? How can it be both waves and particles?

Indeed, in the early stages of quantum theory, such dual, self-contradictory interpretations surfaced many times. This was largely because quantum mechanical concepts are often counter-intuitive; this is necessarily so since our intuition is governed by the macroscopic world, which operates according to the Newtonian laws of motion. Let us look at one such intuitive concept.

In Figure 2.15 we see a ball thrower facing a tall wall. Can he or she throw the ball to the other side? The answer is, 'yes, provided he or she can work up enough energy for the ball to surmount the wall'. If the wall is too tall for the thrower to manage this, the ball will never get to the other side: it will bounce off the wall. This is what the classical mechanics of Newton tells us. In a corresponding problem in the micro-

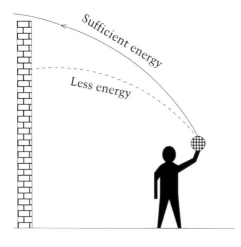

Figure 2.15: The ball thrower must throw the ball with sufficient energy if it is to cross the wall opposite.

scopic world, if we have an electron facing such a barrier, how will it behave? Such a barrier can be created in the path of the electron by other electrons in the vicinity, for example. Such a group of electrons will create an electric field that will repel our incoming electron just as the ball bounces off the wall. In Figure 2.16 we show such a barrier. Considering the barrier as a mountain to climb, we would be tempted to argue from our classical analogy that the electron cannot get to the other side of the mountain if it does not have sufficient energy to cross it. This answer would be wrong! Quantum mechanics holds out another possibility, namely that the electron can tunnel through the

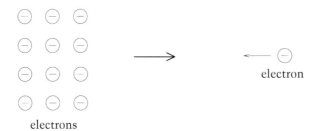

Figure 2.16: The arrow to the right indicates the force repelling back the electron moving to the left. The force thus erects a barrier for the electron. Can it cross the barrier even if it does not have adequate energy to do so according for Newton's laws of motion?

mountain to the other side, even if it does not have sufficient energy to climb and cross it. So the electron may either be turned back by the barrier, or allowed through, there being some finite chance for either of the two alternatives. These odds can be computed by quantum mechanics.

The fact that we cannot definitely assert what the electron *will* do, but can only give odds on what it *may* do, came as a devastating blow to the theoreticians brought up on the deterministic view of Newtonian mechanics. The deterministic view states that given sufficient information about the initial state of a system, and knowing the laws of dynamics, we can tell how the system will behave at any time in the future. For example, the fact that we know in detail about the motions of the Sun, the Earth and the Moon enables us to forecast accurately solar and lunar eclipses in the future. The motion of the electron brought home to the physicists the limitations on the predictability of microscopic systems.

Indeed, the example of the electron is not an isolated one but is generic to quantum mechanics, and this lack of predictability is enshrined in the so-called *uncertainty principle*, which was enunciated by the German physicist Werner Heisenberg in the 1920s, in the early days of the development of quantum mechanics. The wave–particle duality that we find in the behaviour of light is also found in the case of particles. Thus, in the barrier example we in fact argue that the odds on what the electron will do can be calculated by assuming that it behaves like a wave!

Even a great scientist like Albert Einstein, who had been responsible for initiating the concept of the light particle or *photon*, found it hard to accept the uncertainty principle as a fundamental limitation on the deterministic approach. He is credited with the comment: *'God does not play dice.'* He felt that an apparent lack of complete determinism results because the microscopic system may contain other dynamical variables of which an experimentalist working with the sytem is unaware. There were long arguments between him and Neils Bohr, who stressed the fundamental nature of quantum uncertainty. That this argument is still revived from time to time in different forms indicates that many physicists are still not happy with these epistemological issues in quantum mechanics. Nevertheless, all experiments to date looking for the existence of hidden variables have failed and thus lead to a conclusion consistent with the uncertainty principle.

Back to spectral lines

We started on this detour to quantum theory because of the dark spectral lines seen by Fraunhofer. How does the quantum framework explain Fraunhofer's lines?

As an example, imagine a gas of hydrogen atoms lying in the path of solar radiation coming to us. Assume that the gas has all its atoms in the state where their electrons are in one of the inner orbits. Recall that in order for an electron to jump *up* to the next (outer) orbit, it must be supplied with the energy difference between the new and the existing orbit. The solar radiation has photons of varying energies, *including* the specific energy equal to this difference. There is therefore a good chance that the electron will absorb one of the photons of this energy and so jump to the orbit of higher energy. As a result of this absorption, the solar radiation will have a 'hole' at this frequency, which will show up as a dark line against the luminous spectrum.

An atomic physicist can work out the energies an electron can have in all the different orbits in a hydrogen atom. Figure 2.17 shows a typical energy 'ladder'. (The units of energy used in the figure are eV, i.e., electronvolts. One electronvolt is the energy needed to push an electron against an electrical barrier quantified by a potential difference of one volt.) Climbing from one rung to the next means the absorption of photons of a specific energy and hence a specific frequency and wavelength. Recall from Chapter 1 that if we multiply frequency by wavelength we get the speed of light. For example, one such energy difference for a hydrogen atom corresponds to a wavelength of 656 nanometres. What does this mean?

It means that if we examine the solar radiation, we should find it depleted in photons of this wavelength. In other words, we expect to see a dark line at this wavelength. Examining the spectrum of Figure 2.11 we do find a dark line there! It is the line a spectroscopist calls the $H\alpha$ line. Since the wavelength of this line exactly matches the calculated wavelength, the spectroscopist is confident that it has come from the interception and absorption of solar radiation by en-route atoms of hydrogen.

We took the example of hydrogen to illustrate how the method works. There could be, indeed are, other elements that produce absorption in the solar spectrum. The dark lines are therefore called *absorption lines* and from a comparison with theoretical calculations we can fairly confidently deduce the nature and abundance of the chemical

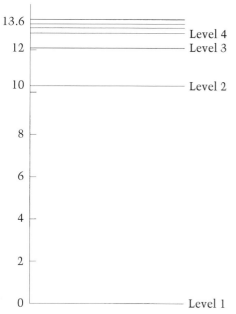

13.6
12
10
8
6
4
2
0

Level 4
Level 3
Level 2
Level 1

Figure 2.17: The 'energy
ladder' of the hydrogen atom.

elements causing them. The identification of a chemical element from its spectral line may be compared to the identification of a criminal from his or her fingerprints!

The abundance part arises from the extent (the heaviness) of the absorption line: the greater the number of absorbing atoms the stronger will be the absorption line. Moreover, as explained below, from the extent of the absorption we can also get a fairly accurate estimate of the temperature of the region where the absorption is taking place. And it is not very difficult to show that these regions are near the outer surface of the Sun. In other words, we now have a diagnostic tool for the surface temperature and surface chemistry of the Sun. It is here that the early theoretical work of Meghnad Saha, an Indian astrophysicist (Figure 2.18), comes in useful.

To appreciate Saha's work, let us first look at what happens when a gas is heated. Normally, a gas consists of atoms or molecules moving around at random, colliding with one another and getting scattered. This internal dynamical activity becomes more and more frenzied and fast as the temperature of the gas rises. Indeed, the temperature is an indicator of the magnitude of energy of this internal motion. So, as we heat the gas, the collisions will be more frequent and more violent,

Figure 2.18:
Meghnad Saha.

leading to the splitting of the molecules into atoms. Furthermore, an atom may lose some of its outermost electrons in a collision. An atom which has been partially or wholly stripped of its electrons is called an *ion*.

During 1918–22, Saha was studying the behaviour of a mixture of hot gas of neutral atoms, electrons and ions. He expected to see in the gas a mixture of some whole atoms, some ions and some free electrons. He also expected to see the proportions of whole atoms diminish and those of ions and electrons rise as the mixture was heated. But how exactly would these ratios change with the rising temperature of the gas? Saha arrived at a formula that gives the exact answer about these relative ratios at any given temperature. Thus from the spectrum we can deduce the abundance ratios and from them the ambient temperature.

It is indeed wonderful that the physics of hot gases coupled with basic ideas in quantum theory can provide us with the means of estimating the surface temperature of the Sun. The same method can of course be applied to the stars, even though they are farther away. Figure 2.19 shows the spectra of some stars with absorption lines of many different elements. Indeed, one discovers, with the help of Saha's formula, that stars with a wide range of surface temperatures exist. These stars have been classified into different spectral classes labelled by O, B, A, F, G, K, M, R, N. Astronomers remember these by a mnemonic originally constructed by Russell, and extended today to be gender-symmetric and to include the last two classes: *Oh Be A Fine Girl/ Guy, Kiss Me Right Now!* The O stars are the hottest (above 30000 degrees Celsius) and contain ionized atoms of helium while the N stars are the coolest (around 3500 degrees Celsius) and contain carbon. A wide variety of chemical elements have been detected in the stars of intermediate classes.

The colour of a star

Another effect of quantum theory provides additional information on the surface temperatures of stars. So far we have taken note of the absorption lines only. But what about the entire continuum spectrum itself? As we mentioned, the visible part of stellar radiation seems to be made of rainbow colours from violet to red; but in what proportion of intensity? If we compare the spectra of two stars, say a hot O star and a cool N star, will we find the different coloured light mixed in the same

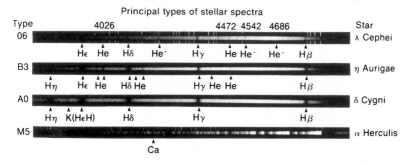

Figure 2.19: Spectra of different types of stars. The absorption lines in these are distributed differently and these help us in estimating the surface temperatures of these stars.

proportion in their spectra? The answer is 'no'. The hotter star will be predominantly *bluer* and the cooler one predominantly *redder*.

This result can now be understood thanks to quantum theory, which tells us how electromagnetic radiation trapped in a limited space distributes itself over different wavelengths. In fact it was the study of radiation in a confined space that led Max Planck (Figure 2.20) to his fundamental idea of quantum theory.

Perhaps a baking oven is the best example of radiation confined to a limited space. Suppose an oven is set at a given temperature and left to heat up. Initially the oven will be supplied with heat from the electric elements or gas flames, which will be hotter than the surroundings. However, as the surroundings receive more and more heat they rise in temperature. Heat flows from a hotter to a colder region and so has the tendency to equalize the temperature everywhere. So, after a few minutes the oven attains its desired temperature, *which is supposed to*

Figure 2.20:
Max Planck.

be the same everywhere. If the oven is well built, its walls will be good insulators and will not allow any significant quantity of heat to be lost.

So here we have the nearest practical demonstration of what a physicist calls *black body radiation*. It is radiation all right: but why *black body*? Because, the radiation is so nicely trapped inside the walls of the enclosure, that to an outside observer, none is detectable. The enclosure with its walls is therefore a black body.

But inside a black body, the radiation shows interesting features should we be able to observe them. For example, the graph showing the distribution of energy of radiation over different wavelengths has a definite shape (see Figure 2.21). Normally, there is very little radiation at low frequencies. For rising frequencies, however, the radiation intensity also goes on rising but only up to a limit, beyond which it sharply declines. Moreover, the shape of the graph can be calculated using

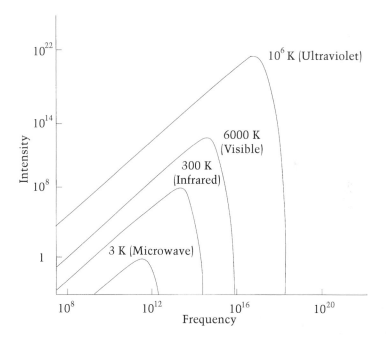

Figure 2.21: These curves show how the intensity of radiation in a black body rises and falls with its frequency. Each curve corresponds to a black body of fixed temperature. The frequency at which the intensity peaks increases in proportion as the temperature of the black body is raised. The typical nature of the radiation is given in parentheses.

quantum theory and it is completely specified by the temperature of the radiation.

In the figure we see different black body distribution curves for various different temperatures. Notice that the higher the temperature, the higher is the distribution curve. Also, at higher temperatures the peak of the curve shifts to the right, that is, towards higher frequency. As shown in Figure 2.21, the radiation will appear predominantly as microwaves at low temperatures and as X-rays at high temperatures. We will return to this feature shortly.

We pause here to discuss the scale for measuring temperature. We talk of a temperature of 98.6 °F as normal for the human body, or of 100 °C as the boiling point of water. The scales Fahrenheit (°F) or Celsius (°C) are used for convenience as well as for historical reasons. To the physicist, however, the natural scale for temperature is the *absolute* scale, which measures the internal energy of a body. This energy arises from the motions, rotations, vibrations etc. of its constituent atoms and molecules. As the temperature rises this internal energy rises. Conversely, if we cool the body these internal motions slow down. The state when they come to a complete stop is the state of zero temperature, *provided that temperature is measured on the absolute scale*. On the Celsius scale this corresponds to a temperature of 273 degrees *below zero*. We shall use the absolute scale frequently, as it is natural for our discussion. A temperature on this scale is denoted by the letter K for the physicist Lord Kelvin (Figure 2.22), who played a key role in the early development of these concepts. Thus −273 °C = 0 K.

We return now to the black body distribution curve. In the days before quantum theory, physicists had tried to understand this distribution using the classical theory of electromagnetic radiation. They had only partial success. They assumed, of course, that the radiation inside the black body comprised light of different wavelengths. They then tried to work out how the available energy would be shared by the different wavelengths after the initial give and take has led to a steady situation. They found that they could reproduce the left-hand part of the curve but not the right-hand part: thus classical theory predicted that the intensity would continue to climb with frequency, and would never drop! And this led to the absurd situation that a black body was predicted to emit infinite energy – known as the ultraviolet catastrophe.

However, with Planck's assumption that light is not just made of waves but is also distributed in tiny packets of energy (quanta), it became possible to arrive at the observed distribution. Indeed one can

Figure 2.22:
Lord Kelvin.

estimate the value of Planck's constant h (see earlier in the section) from these studies.

But, how, you may ask, can one observe what is going on within a black body. Isn't its interior sealed off from the outside observer? That comment is indeed correct and one needs to compromise a bit! Suppose we puncture a few tiny holes in the walls and collect the escaping radiation. Provided we make the holes small, the amount escaping will hardly be noticed inside because the state of equilibrium there would not be significantly disturbed. And yet the escaping radiation can be examined and it will give us an indication of the state within.

In our analogy of the heated oven, if we open the oven door to find out how hot it is, its radiation will escape, thus destroying the state of equilibrium. Instead, a clever device for measuring its temperature ensures that the measuring process does not destroy the state of equilibrium.

These experimental studies of black body radiation by the pre-quantum physicists had led to an interesting result about the distribution peaks in Figure 2.21: the peak frequency is in strict proportion to the temperature of the black body. This law is named after W. Wien, who discovered it in 1894. The temperature here is measured not on the Celsius or Fahrenheit scale, but on the absolute scale.

For example, if the black body has a temperature of 3 K, the peak frequency will occur at three hundred thousand million cycles per second, while at ten times this temperature the peak frequency will also be ten times greater, i.e., three million million cycles per second.

So we see that high peak frequency corresponds to high temperature. Thus if we recall from Chapter 1 that blue has a higher frequency than red, then we will likewise find that the blue intensity will dominate over the red intensity at relatively high temperatures and vice versa for low temperatures.

How does all this concern stars? It does, because stars do approximate closely to black bodies. This may appear contradictory: how can a shining object be like a black body? Recall, however, our analogy of the oven with a few tiny holes. So long as the leakage of radiation from the oven is small enough not to disturb its internal equilibrium the black body approximation is a reasonable one. In the case of the shining star, the flow of radiation coming out from its surface is not high enough to disturb its state of equilibrium in the layers underneath. And so we can fit the continuum spectrum of the star with a black body curve and estimate its temperature. This agrees quite well with what we got earlier from the absorption line spectra. We also see why blue stars are hotter than red ones.

STELLAR SIZES

The fact that stars radiate as black bodies, at least approximately, enables the astronomer to estimate their sizes. This is illustrated in Figure 2.23. This diagram shows the amount of radiation coming from black body spheres of different radii but the same temperature. Each solid line in this diagram corresponds to one radius. As we move along the line towards the right we encounter stars with larger temperatures and larger luminosities. As the radius is increased, we move up to a higher line. Thus two stars with the same radii but different surface temperatures will radiate differently: the star with the higher surface temperature will be more luminous.

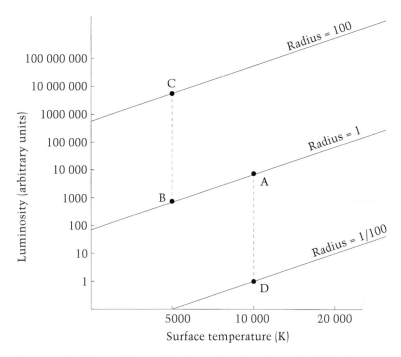

Figure 2.23: Each diagonal line in this figure shows how the total radiation from black bodies having the same radius but different temperatures depends upon the radius. The different lines show how this radiation pattern changes with radius.

Just for a comparison, consider three stars, *A*, *B* and *C*, in Figure 2.23. Stars *A* and *B* are of the same radius, but the temperature of *A* is twice that of *B*. Then *A* will be sixteen times as luminous as *B*. Star *C*, however, has the same surface temperature as star *B*, but has a radius a hundred times larger. Then *C* will be ten thousand times as luminous as *B*.

Now look at stars *A* and *B*. On the one hand, star *A* is much more luminous than *B*, the reason being that its surface temperature is much higher. On the other hand, star *D* has the same temperature as star *A*, but much lower luminosity. Why? By an argument similar to that for the pair *C* and *B*, we conclude that the radius of star *D* will be about a hundredth of the radius of *A*.

We thus have a comparative statement about the stellar radii of stars *A*, *B*, *C*, *D*: *Star D is about a hundredth of the size of star A; stars A and B are comparable in size, but star C is hundred times as large as A or B.*

We now look at the H–R diagram, reproduced here as Figure 2.24, and compare it (broadly) with the mirror image of Figure 2.23. If we call the main sequence stars *A* and *B normal* stars then star *D*, being much smaller than normal, is called a *dwarf* star while star *C*, being much larger than normal, is called a *giant*.

Thus we have a wide variety of sizes of star in the stellar world, a much wider variety than found in the human population. From the tiniest of new-born babies to the tallest adult, the span of human heights is no more than over a factor 5. For stars, the range of radii from dwarfs to giants extends over a factor of more than 10 000.

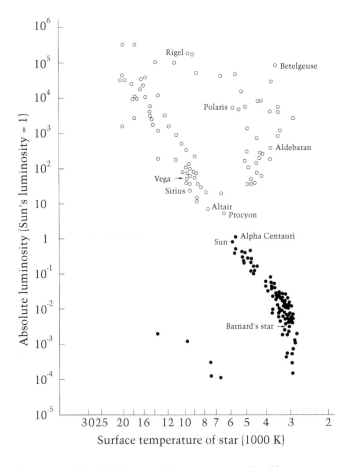

Figure 2.24: The H–R diagram of Figure 2.4 is reproduced here.

The H–R diagram therefore gives us a clear indication of the range of variation amongst the stellar population. It does not, of course, answer the question that we now would like to ask: how does this variety come about? Are stars born as we find them on the H–R diagram or does a typical star pass through many different states as it evolves, including states when it is 'normal', when it is a 'giant' and when it is a 'dwarf'? The remarkable progress made by stellar physicists during this century enables this question to be answered clearly and succinctly. As we shall now see, the clue lies in the answer to the basic question: *What makes a star shine?*

THE SECRET OF SOLAR ENERGY

The above question is perhaps one of the oldest that has stimulated human curiosity. Certainly, so far as the Sun is concerned, considering its enormous radiating power it is not surprising that the ancients attributed divinity to this heavenly body. The huge Sun temple in Konarak on the eastern coast of India (see Figure 2.25) bears testimony to such beliefs.

With the rise of modern science in the seventeenth century, how-ever, the mechanistic view began to prevail. The belief that natural phenomena should ultimately find an explanation in terms of a few basic laws of science extended to astronomy also. In particular, the *law of conservation of energy* began to acquire a universal status.

This law states that in any process the total energy involved is always conserved: it is neither created nor destroyed.

So, to apply it to the Sun would mean that there is a source within it from which energy is being acquired, a source that would surely deplete with time. What is this source?

Two distinguished physicists of the nineteenth century, Baron von Helmholtz of Germany (Figure 2.26) and Lord Kelvin (Figure 2.22) of Britain, whose name we have encountered in the context of the abso-lute scale of temperature, attempted an answer to this question. Their solution appealed to the reservoir of gravitational energy that any mas-sive body possesses. Let us digress a little to find what this reservoir is.

In Chapter 5 we shall explore the manifold wonders associated with the phenomenon that we call *gravitation*. Here we will limit ourselves to the very basic feature of the gravitational force as outlined by Isaac Newton in his *law of gravitation* in the seventeenth century. The law is

Figure 2.25: The Sun Temple at Konarak in Eastern India depicts the Sun god riding in a huge chariot.

simple to state but, as we will see in Chapter 5, it has profound implications. The law states that *any two material bodies attract each other with a force which is in direct proportion to the material content of each of the bodies but whose strength falls off in inverse square proportion to the distance between them.*

The material content of a body is measured by its *mass*. In our day-to-day life we use the *kilogram* unit to measure mass. Suppose we have two bodies A and B each of mass one kilogram, separated by a distance of, say, one metre. According to Newton's law of gravitation, there will be a certain force of attraction between A and B. If we replace A by another body C, say of mass 10 kilograms, the force of attraction between C and B will be 10 times what it was between A and B. Likewise, if we increase the distance between A and B to 10 metres, the force of attraction will reduce by a factor 10 × 10, that is, by a factor *one hundred*. (Recall that we have already encountered the concept of inverse square proportion in the context of luminosity.)

Let us apply the law of gravitation to a massive spherical body such as the Sun. In Figure 2.27 we see two typical parts A and B of this body.

Figure 2.26:
Baron von Helmholtz.

According to the law of gravitation, the two parts will attract each other
and so they will tend to approach each other along the straight line
joining them. But recall that A and B are typical parts; and as such,
the same rule will apply to any other two parts of the Sun. The result
is that there are internal pulls within the Sun tending to shrink it to a
smaller size. In Chapter 5 we will have occasion to elaborate on this
tendency of massive bodies.

This same tendency contributes a gravitational energy reservoir to
the Sun. Physicists express the tendency towards inward motion by
saying that the reservoir has *potential energy*. We encounter this type
of reservoir in a gravity dam, such as shown in Figure 2.28. In such a
dam, there is a water reservoir at a height and water is released from
there to fall downward. Because of the inward force of gravitation
towards the centre of the Earth, this falling water acquires speed and
can therefore be used to drive turbines for generating electricity in a
hydroelectric power station.

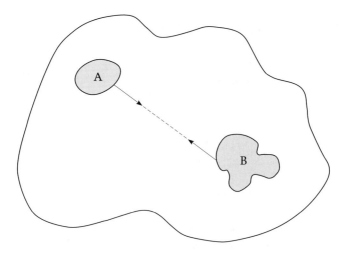

Figure 2.27: Any two parts *A* and *B* of a massive body will attract each other and will thus have a tendency to move towards each other along the line joining them. The overall tendency of all such gravitational forces is to make the body contract.

Figure 2.28: The dam at Bhakra Nangal in the Punjab is one of the world's tallest gravity dams.

The gravity dam is an excellent example of the transformation and conservation of energy. The original energy of the water is gravitational, by virtue of its high position; it turns into energy of motion as the water falls and eventually the energy of motion is converted into electrical energy. However, the total energy remains the same, only it changes its form.

In the same way, in a spherical mass there resides gravitational energy, which can be tapped by contracting the sphere. Kelvin and Helmholtz believed that the energy radiated by the Sun comes from this reservoir. Imagine the past history of the Sun, when it was considerably more extended and diffuse. By the above tendency of gravitation to contract the extended sphere to the Sun's present size, energy is released. The energy could be estimated and it was found sufficient to have kept the Sun shining for some 20 million years.

Unfortunately, it turns out that while 20 million years is quite long compared to the lifespan of human civilization, it is by no means long enough for the Sun. For, from the radioactive dating of the ages of meteorites and terrestrial rocks one finds that the solar system has an age of approximately 5 *billion* years. This is evidence that the Sun has been steadily shining at its present rate for a period of this order. Clearly the Kelvin–Helmholtz prescription is not sufficient for meeting this demand.

Thus by the 1920s, the problem was back on the drawing boards of theoreticians: to find a source of energy large enough to keep the Sun shining at its present rate for at least 5 billion years.

The Cambridge astrophysicist Arthur Stanley Eddington (Figure 2.29) arrived at the correct solution through his investigations of the internal constitution of the Sun. Eddington imagined the Sun to be a hot ball of gas held together by its own gravitational force, as imagined above. He then set up a system of equations which relate to the internal structure of the star. We will not go into the technical details here but will nevertheless highlight the argument that led him to the mysterious source of solar energy.

To understand the argument let us look at an every-day example, a deep-sea diver who penetrates to great depths under the sea. One effect that the diver experiences is a rise in pressure with increasing depth below sea level. At a depth of about 10 metres the pressure is double that at sea level. And it goes on increasing at the same rate, that is, it will triple at 20 metres, quadruple at 30 metres and so on. Why?

Figure 2.29:
A.S. Eddington.

The pressure at sea level is the result of the column of atmospheric air that the Earth's surface has to support. Just as we humans experience our weight because all of us are gravitationally attracted towards the Earth, so does even the tenuous air above us have weight. And pressure is simply this weight per unit area of the Earth's surface. The pressure exerted by the atmosphere amounts, approximately, to the weight of ten thousand kilograms distributed over a square surface of one metre by one metre. (See Figure 2.30 (a).)

Our deep-sea diver has to bear not only the weight of this air column but also the weight of water above. And this latter weight increases as the diver dives deeper. But what has all this to do with a star like the Sun?

As illustrated in Figure 2.30(b), the pressure inside a star rises as we go deeper inwards, for the same reason that the pressure in the sea rises with depth. The difference between the sea and the star is that the star is made of gaseous material while the sea is liquid. One of Eddington's equations tells us how the pressure inside the gas is related to its temperature and density.

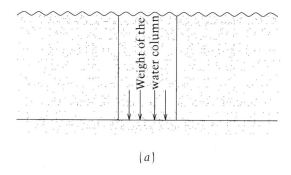

(a)

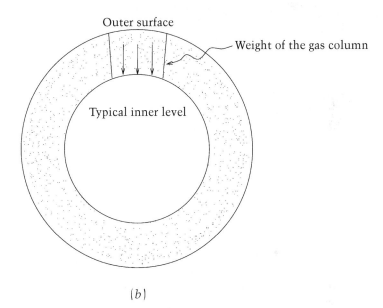

Outer surface

Weight of the gas column

Typical inner level

(b)

Figure 2.30: In (a) we see that deep under the sea the pressure mounts because any horizontal surface at that level has to support the weight of the column of water above it. In (b) a similar situation is seen to exist within a star, with pressure mounting towards the centre.

There is another difference between the star and the sea. As Eddington discovered, within the star there is a huge store of radiation and radiation has its own pressure. The toy shown in Figure 2.31 demonstrates that even the radiation from an electric bulb exerts pressure. The paddles reflect light on one side and absorb it on the other. The former process imparts a greater push on the panel than the latter

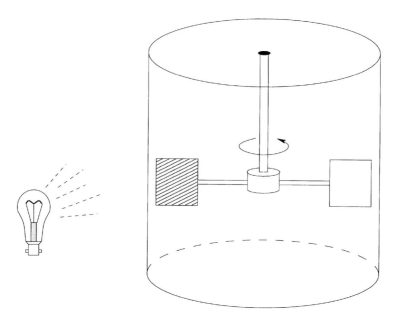

Figure 2.31: Light shone on this thin-leafed toy would make it rotate because of the pressure exerted by the radiation on the metal leaves.

and the net pressure of radiation moves the panels. Likewise, one has to add the radiation pressure to the gas pressure in the deep interior of the Sun. And both kinds of pressure lead to the conclusion that as the pressure rises inwards, so does the temperature.

Note that the temperature at the surface of the Sun is around 5750 K. So the temperature at the centre will be higher. Eddington's calculations gave the remarkable answer that for a star like the Sun, the central temperature would be in excess of *10 million degrees*. Never before had anyone come up with a temperature of a physical object as high as this!

However, in this entire picture of the Sun's interior, one item was missing. What it was that kept the core of the Sun at such a high temperature and supplied the energy that the Sun was radiating out? Evidently, the calculations pointed to a source of energy in the central region that did both.

Recalling a suggestion of J. Perrin, Eddington now suggested how the Sun had managed to produce so much energy for so long. His argument was briefly as follows.

The lightest atom in nature is the atom of hydrogen, consisting of a proton and an electron. At the high temperatures prevailing in the Sun, however, these atoms would not be able to retain their structure: they would lose their electrons through the highly energetic collisions that take place frequently. Thus there would be the nuclei of these atoms roaming around freely along with a sea of electrons. Such a state of matter in which the atomic electrons are dissociated from the nuclei is called the *plasma* state (see Figure 2.32).

The stable nucleus next higher in mass is that of helium. It carries two protons and in addition, two electrically neutral particles called *neutrons*. The mass of a helium nucleus is found to be a little less

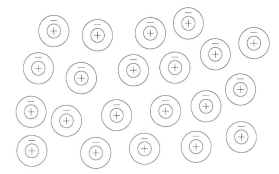

(*a*) Particles of gas at moderate temperatures

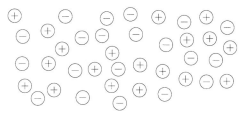

(*b*) Plasma at high temperature

Figure 2.32: At high temperatures an atom may lose some or all of its electrons and be left as a positively charged *ion*. A combination of electrons and ions makes up the plasma state of matter. In (*a*) we see a collection of neutral atoms of a gas at moderate temperature. In (*b*) we see the atoms split up and the gas changed to the plasma state.

than the combined mass of four hydrogen nuclei. Now, argued Eddington, suppose four nuclei of atoms of hydrogen combine in a nuclear process and get converted to an atom of helium. What has happened to the lost mass, the mass deficit? The law of equivalence of mass and energy typified by Einstein's famous equation $E = Mc^2$ tells us that the mass deficit will manifest itself as energy. *This is the energy that is available to the Sun to radiate.*

The energy so available is a rather tiny fraction of the energy equivalent of the mass of four hydrogen atoms. In fact, modern calculations show that only 7 parts in 1000 are available for radiation. Yet this reservoir is so vast that it has not only lasted the Sun for five billion years but will be good for another six billion years. (In terrestrial terms, we get some idea of the vastness of this energy source from the calculation that one kilogram of hydrogen fuel used in a fusion reaction can keep a megawatt generator running continuously for about 20 years.)

However, in the twenties nuclear physics was a nascent science. The nature of the force that binds the neutrons and the protons into the nucleus was not known. To the atomic physicists of his times, Eddington's ideas appeared outlandish.

We can see one difficulty, for example. We know that like electric charges repel each other and the force of repulsion grows in inverse proportion to the square of the distance separating them. (Note that again we have the inverse square law: but this time, unlike gravitation, we are talking about a force of repulsion.) So how can nuclei of hydrogen atoms, which are positively charged protons, come close enough to stick together to form a helium nucleus?

Eddington's argument was that, at the very high temperatures in the core of the Sun, the protons would be moving so fast that it was not unlikely that two of them could surmount the barrier erected by the force of repulsion and come close enough to be fused by the hitherto unknown nuclear force. The atomic physicists disagreed with this conclusion, feeling that the temperature would not be high enough to promote such a reaction.

In his classic textbook *Internal Constitution of the Stars* written in the early 1920s Eddington has the following reply for his atomic-physicist critics:

> We do not argue with the critic who tells us that the stars are not hot enough for this purpose. We tell him to go and find a hotter place . . .

In the end, the controversy was resolved in Eddington's favour. By the late 1930s, nuclear physics had progressed to the extent that the nature of nuclear binding was becoming understood. The attractive force between nuclear particles, the protons and the neutrons, operates irrespective of whether the particles are electrically charged or not. Moreover, the force is of very short range: it ceases to operate beyond a range of around a thousand million millionth part of a metre. But within that range, it is so strong that it completely dominates the electrical repulsion between the protons in the nucleus.

So at high temperatures exceeding ten million degrees, two protons *could* come close enough to be trapped by the nuclear force, and through such fusion in stages, the larger nucleus of helium can be made in the Sun's core. In 1938–9 Hans Bethe (Figure 2.33), a nuclear physicist, was able to use this information to construct a complete model of the Sun.

The proof of the pudding

To the lay person, all this may be interesting but speculative. How do we know that such nuclear fusion is indeed going on inside the Sun?

Figure 2.33:
Hans Bethe.

How can we ascertain whether the model of the Sun so constructed is reasonably correct?

In science too such questions are inevitable. A scientific theory has to be tested by observations before being passed as reasonable. In the case of Eddington's framework, the theory gave a unique relationship between the mass and luminosity of a star. The higher the mass, the higher the luminosity. Likewise, the theory predicted a mass–radius relation. Now in the case of the Sun, astronomers can estimate the mass of the Sun from the gravitational pull it exerts on the Earth and other planets. So we can predict the luminosity and radius of the Sun and place it on the H–R diagram, *from purely theoretical considerations*. And then we can compare it with the position obtained observationally. The two match very well. Not only that; if we carry out this theoretical exercise for stars of other masses, both greater and less than Sun's mass, we will get a theoretical curve on the H–R diagram of the kind shown in Figure 2.34. A comparison with Figure 2.4 tells us that this curve is nothing other than the main sequence of the H–R diagram.

So we have not only a demonstration of the correctness of the theory but also the answer to why we find stars on the main sequence. But can we be more ambitious and seek further confirmation of the theory? In particular, is there any way we can actually measure the central temperature of the Sun? This may appear to be out of the question. Not only is the Sun's interior inaccessible to us, but it is also impossible to see. The whole body of the Sun forms an opaque sphere which prevents us from seeing what goes on within. Even so, scientists have found a way round the difficulty. We will defer this remarkable exercise to the Epilogue, for it generates a new puzzle that has still to be resolved.

The Sun's example points a way to a solution of the energy problem that is now facing the human race. The petrochemical resources of this planet are limited and may not last long. Some say that they will be adequate for a few centuries, while others, more pessimistically, say that they will be exhausted within decades. So we need to look to other resources for our energy needs. Can we carry out the process that the Sun has been operating for so long in a laboratory on Earth? Known as *thermonuclear fusion* (the fusion of atomic nuclei at high temperatures), this process is indeed being attempted in the laboratory. An explosive version of the process has already been achieved, unfortunately: the hydrogen bomb, with enormous destructive potential. What is needed is its controlled version. We have to find a way of creating a steady output of energy as the Sun has managed to do.

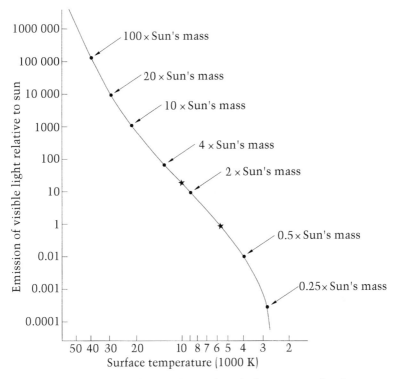

Figure 2.34: This theoretical curve, showing how the luminosity and surface temperature vary for stars of different masses, may be compared with the observed main sequence of the H–R diagram of Figure 2.4.

In this case the Sun does have one enormous asset that we humans lack. Because of its huge mass it produces large gravitational pressures to hold the hot plasma at the centre in steady equilibrium. Without this enormously strong gravity to fall back on, the test of human ingenuity lies in creating an alternative scenario which provides a hot but stable plasma. When achieved, it could well be a modern wonder of science and technology.

But we now go back to the Sun's own energy problem: with the thermonuclear energy that is available in the above fashion, how long will it last? As mentioned before, calculations show that the energy reservoir of the Sun is adequate not only for its life to date of around five billion years, *but also for another six billion years to follow.* The time for which a star will be able to tap its hydrogen reservoir depends on its mass. The more massive stars have shorter times while the less massive ones go on for a longer duration.

RED GIANTS

Long though this span is, it is legitimate to ask: What will happen to a star that has no more hydrogen left to fuse into helium?

Figure 2.35 illustrates such a star. It has an inner core made of helium and an outer envelope made of hydrogen. We recall that the temperature of the star is over ten million degrees in the core and drops down to a few thousand degrees at the surface of the envelope. Thus, although there is hydrogen in the star, it is much too cool to be fused into helium, which is why energy production in the star has come to a halt.

In the absence of energy coming out from the centre, the core is no longer able to hold out against its inward pull of gravity. For, in an energy-producing star the huge pressures of radiation and hot matter successfully combat the inward pull of gravity. Once energy production has stopped, these pressures became inadequate to keep the core intact against gravitational contraction. The core therefore contracts.

In general, when a gaseous mass contracts it tends to heat up. The temperature of the core therefore rises as it shrinks. And as the temperature reaches close to the hundred million mark, a new fusion reaction is triggered off within it, a reaction which will now generate energy

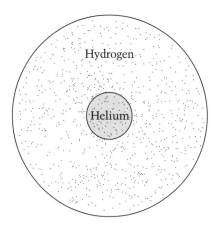

Figure 2.35: When the star has finished the fusion of all its fusible hydrogen it has a helium core and an outer envelope predominantly containing hydrogen at a lower temperature.

for the star. What could be that reaction? Could it build even bigger atomic nuclei from the building blocks of hydrogen and helium?

A historical interlude

In the 1950s, several physicists were grappling with this problem. Studies of nuclear structure suggested the prima facie possibility that the fusion process could, in principle, continue towards the building of bigger nuclei. However, specific details prevented progress. The difficulty can be imagined from the following analogy.

Suppose you are erecting a boundary wall by placing layers of stones one on top of another. However, beyond a certain height the wall becomes unstable and all the layers collapse. How then do you proceed at all?

The problem with the fusion of nuclei was that having made the nuclei of helium, the next step would involve putting together either two nuclei of helium or a combination of a helium and a hydrogen nucleus. In either case, the resulting combination would be an unstable nucleus which broke back into smaller ones.

The problem was solved in a bold fashion by the Cambridge astrophysicist Fred Hoyle (Figure 2.36). Hoyle argued that instead of looking for fusion of two nuclei, why not have a fusion of three? (In the stone wall analogy, placing one stone on top of another may not give a stable structure, but intermeshing three stones together may prove successful.) Hoyle suggested that three nuclei of helium may fuse together to provide a stable nucleus, that of carbon.

In fact this possibility had occurred earlier to others also, but there had been a difficulty that seemed insurmountable. Recall that, in a hot gas, fusion of three helium nuclei would take place provided all three of them arrived in the same place at the same time. Since they are moving in random directions, the chance of this occurring would be rather small. So a process based on such rare events would proceed very slowly, unless some way could be found to compensate for its slowness.

This is where Hoyle found a solution. He suggested that to compensate for the rarity of such a three-body collision, the fusion process must involve a *resonant* reaction. What is a reasonant reaction?

Resonance is known to us in sound. When a violinist tunes the strings of his instrument to the right tension, it resonates to certain notes. That is, the frequencies of vibration of the strings match the vibration of the air in the hollow of the instrument. The result is ampli-

Figure 2.36: B²FH: Margaret Burbidge, Geoffrey Burbidge and William Fowler with Fred Hoyle.

fication of those notes. This exact matching is called resonance. It goes beyond the above example of sound, of course, to encompass other phenomena where frequency matching takes place.

In a resonant nuclear reaction the energy of the three participating nuclei should exactly match the energy of the new carbon nucleus formed. In such a situation the reaction is very likely (just as the sound of the violin note is amplified). This high probability compensates for the rarity of a three-body encounter. Unless such a resonance is present, argued Hoyle, there will not be any significant production of carbon in the star. Or, to put it the other way round, in order for the star to have a continuing source of energy through fusion *it is essential that such a state of resonance exists.*

Armed with this argument, Hoyle, who happened to be visiting California Institute of Technology in 1954, asked the nuclear physicists to verify whether such a state of energy existed for the carbon nucleus. He expected this energy to be somewhat *higher* than the state of energy for a standard nucleus of carbon. In the jargon of nuclear physics, such a nucleus is said to be in an *excited* state. The excited state does not last long, however, and the nucleus returns to the standard state by releasing the extra energy. This is the energy that the star could draw upon in order to continue shining.

The nuclear physicists were sceptical about this entire chain of arguments. (Recall the earlier encounter of Eddington with the atomic physicists!) Nevertheless, Ward Whaling and Willy Fowler and others at the Kellogg Radiation Lab at the California Institute of Technology decided to check this apparently outlandish prediction from an astrophysicist. And they found that Hoyle was right: that an excited state of the carbon nucleus did exist, exactly as he had expected.

As we shall discuss in the following chapter, Fred Hoyle had another motivation for arriving at this remarkable prediction, perhaps more compelling than the requirement that the star should continue shining even after all fusible hydrogen has been used up. For the time being, however, let us follow the progress of the star.

When the stellar core becomes hot enough, say at a temperature of a hundred million degrees, the helium nuclei which hitherto were lying inert begin to take part in a new fusion reaction. A set of three helium nuclei can combine together to form a carbon nucleus in a resonant reaction. The carbon nucleus is in an excited state and it decays to the standard rate releasing the extra energy (see Figure 2.37). Let us see how all this affects the overall structure of the star.

The formation of a red giant

The activation of a new source of energy results in the rebuilding of the pressures within the core, which now ceases to contract. These pressures are thus able to cope with the inward pull of gravitation of the

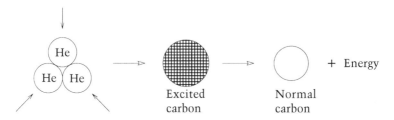

Figure 2.37: In the process suggested by Hoyle, three nuclei of helium are fused together to form an excited state of a carbon nucleus (shown by the hatched sphere), which decays into a standard nucleus of carbon while releasing some energy.

core. However, the increase of pressure cannot be confined to the core only. In order to adjust to the new situation, the envelope also develops increasing pressures which begin to blow it outwards. The envelope therefore expands steadily and will settle down to a new size which may easily be more than a hundred times the original size of the star. The rate of energy output from the star also picks up, that is, it becomes more luminous.

However, just as the core heated up due to contraction, the envelope cools due to expansion: its outer surface temperature may drop by a couple of thousand degrees or more. Remembering our discussion of how the temperature of the stellar surface is related to its colour, a golden star will turn reddish on expansion.

This is our *red giant*. The Sun will become one after it has exhausted its fusible hydrogen fuel. In that state it may become so large that it would certainly swallow the inner planets of Mercury, Venus and Earth and most likely Mars also.

What will happen to the inhabitants of this planet as it gets swallowed by the Sun? Let us hope that they will be so advanced that they may abandon the planet well in time before the going gets rough. They may prefer to reside on or near one of the moons of Jupiter. Anyway we need not worry about this right now as the event lies some six billion years in the future!

Perhaps we may appreciate the story of the professor who was explaining all this to his student over a glass of beer in a pub. A person who was sitting at a neighbouring table, and who had had one glass too many, staggered to them with a worried face: 'Professor, did I hear you say that the Sun would swallow the Earth in six million years?', he asked. 'No, Sir, not six million but six billion years is what I said', replied the Profesor. The drunk heaved a sigh of relief and said: 'Ah, then I don't need to worry'.

FROM GIANTS TO DWARFS

We thus have a theory that explains the red giant star as the next stage in the evolution of a star after it has finished the hydrogen fuel. The star then becomes larger in size, cooler at its outer surface but more luminous. It therefore moves in the H–R diagram off the main sequence towards the right and top, which is where the giants are. Our next question is: How are dwarf stars formed?

The evolutionary scenario after the red giant stage will be discussed in the next chapter. However, we can see the dwarfs as one possible ending to that scenario. It comes about when a star has no more nuclear fuel *of any kind* left to burn. Since any fuel reservoir will be exhausted sooner or later, it is a legitimate question to ask what happens to a star at that stage.

As we would expect, there will then be no significant opposition to gravitational contraction of the star: in the absence of any energy generation, the pressure forces will begin to fall far short of what are needed to withstand the force of gravitation. But will gravitation be allowed to dominate the situation in every case?

The answer is 'No'. For, imagine that the material in a given volume is compressed indefinitely. Its density will rise and there will come a time when all the atoms in it are very tightly packed. There is a new restriction, of a quantum mechanical nature, that intervenes at this stage. This restriction becomes relevant for a system containing several identical particles of matter of the fermion type. (Fermions are identical particles that have spin $1/2$, $3/2$, ... Prime examples are electrons and neutrons.) In the case of the white dwarf stars these particles are electrons.

We recall that in stars the atoms are usually in plasma form, with positively charged ions separated from the negatively charged electrons. The state of a typical electron is specified by its energy, momentum and spin, and the new quantum mechanical rule that becomes operative is that we cannot find two such electrons in the same state, with exactly the same energy, momentum and spin direction. Since the number of available states for the electrons of any given energy is limited, because the rungs of the energy ladder become more widely separated as the star shrinks, the electrons in a high-density material will resist close packing beyond an admissible limit. Electrons which have reached such a limit are said to have become *degenerate*.

The above rule is called *Pauli's exclusion principle*, after the quantum physicist Wolfgang Pauli. It leads to the building up of new pressures, called *degeneracy pressures*. It is these pressures that call a halt to any further shrinking of the star.

The Chandrasekhar limit

In the mid-1920s, the Cambridge physicist Ralph Howard Fowler had used this result to work out the equilibrium states of very dense stars that have no nuclear fuel left for burning. In such stars the degeneracy pressure holds back the contracting tendency of gravitation. Fowler found that in this way stars of *any* mass could be supported in states of equilibrium. Such stars would emit very feeble radiation, drawn from their gravitational reservoir, as in the Kelvin–Helmholtz hypothesis discussed earlier for the Sun. That is, the stars would shrink, but very slowly, and use the gravitational energy released in the process to emit their faint light. These would be the white dwarf stars.

Thus, it was felt that the problem of white dwarfs was sorted out. But no! There was more to come.

In 1930, Subrahmanyam Chandrasekhar (Figure 2.38), a young Indian from Madras, began to think about this problem while on board the steamer to England where he was going for a research degree.

Figure 2.38:
S. Chandrasekhar.

He found one lacuna in Fowler's argument. The problem can be understood with an analogy illustrated in Figure 2.39.

Here we see a bucket being filled with water. Since the bucket has a limited cross section, as more and more water is poured in its level rises. In the case of a white dwarf star, the compression of matter together with Pauli's principle tells us that only a limited number of electrons can be accommodated in any given volume up to a specified level of energy. If more electrons have to be accommodated in a given volume, as will happen if the star continues to shrink, their energy levels must rise, just like that of water in the vessel. Furthermore, as their energy rises the electrons begin to move faster. And Chandrasekhar could work out that for more massive stars the velocities may approach the speed of light.

Now, in 1905, Albert Einstein had already demonstrated that the notions of space and time measurement need to be revised to keep in conformity with the observed phenomena in electricity and magnetism. Consequently, the laws of motion also need to be revised from the formalism given by Newton in the seventeenth century. The new rules came to be known as the special theory of relativity. (See Chapter 5 for details of this theory.) Modifications from the Newtonian laws of motion become significant for objects moving with speeds approaching that of light. So, argued Chandrasekhar, for more massive stars we should use the special theory of relativity rather than the Newtonian laws of motion.

Chandrasekhar worked on this problem and soon found that the dependence of the degeneracy pressure on the density of matter is altered in the relativistic regime: larger stars are 'softer'. Because of this, Fowler's earlier result, based on Newtonian ideas, had to be modified. In particular, Chandrasekhar found that there is a limit on the mass of a star, above which it cannot be supported by invoking degeneracy pressures. This mass limit is 1.4 times the mass of the Sun. *That is, stars with masses more than 40 per cent higher than the solar mass cannot exist as white dwarfs.*

This was indeed a remarkable result and it very strikingly demonstrated how the rules of the microworld can determine the properties of such huge objects as stars. Yet, when Chandrasekhar presented his conclusions to the august gathering of astronomers at the Royal Astronomical Society on its traditional second-Friday-of-the-month meeting on 10 January 1935, he rather unexpectedly encountered a hostile reception from no less a person than Eddington himself.

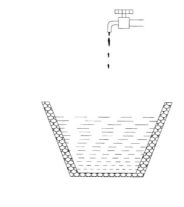

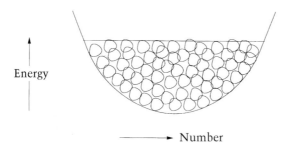

Energy

Number

Figure 2.39: In the upper figure we see how the water level in a bucket rises as it is fed with more and more water. In a dense star, contraction increases the density of electrons and this leads to their occupying higher and higher levels of energy as they fill up all the available but limited space.

What disturbed Eddington about Chandrasekhar's result was an important consequence of this critical mass limit on white dwarf stars. While one could rest assured that stars below this limit would continue to exist as white dwarfs, what happens to those whose masses lie *above* this limit? What is the future of such a star if it cannot have degeneracy pressures within? Such a star would go on contracting and emitting radiation, but what would be the endpoint of the process? Eddington argued:

> . . . The star has to go on radiating and radiating and contracting and contracting until, I suppose, it gets to a few kilometres radius, when gravity becomes strong enough to hold in the radiation, and the star can

at last find peace . . .. Various accidents may intervene to save a star, but I want more protection than that. I think there should be a law of nature to prevent a star from behaving in this absurd way!

Eddington therefore felt that the arguments employed by Chandrasekhar to arrive at such an 'absurd' conclusion had to be wrong. And the weight of his authority and personality went a long way towards creating the impression at the RAS meeting that he was right.

Nevertheless, Chadrasekhar was proved eventually to be correct and the limit on white dwarf mass that he had derived came to be known as the *Chandrasekhar limit*. White dwarfs must have masses *below* this limit. Observations so far bear out this conclusion.

But what about those stars that are unfortunate enough to exceed this upper limit on their mass? Ironically, here Eddington had been right in his forebodings as to their future, but his expectations as to what nature should do were not borne out. For, nature has the habit of transcending human expectations, leading to results more dramatic than the human mind can imagine.

Had Eddington taken Chandrasekhar's results seriously he would have been credited with the prediction of *black holes*. But, more of this story in Chapter 6.

When stars explode . . . ③

AN EVENT THAT SPANNED CENTURIES

What is common to the following: a Chinese emperor of the Sung Dynasty, a learned physician from the Middle East, the Red Indian tribes on the American subcontinent, all belonging to the eleventh century, and the astronomers of the twentieth century?

Sounds like a quiz question? It could very well be!

A cryptic anwer is: They were all witnesses to a spectacular cosmic event, which is still unfolding, an event that was first witnessed here on Earth on 4 July 1054. But its aftermath is still being studied today, and will continue to be investigated by astronomers in the years to come.

The event, and others like it, very well qualifies for inclusion in our list of cosmic wonders.

Let us begin with the Chinese, to whom we are indebted for maintaining records that date back nine and a half centuries.

THE GUEST STAR

In the *History of the Sung Dynasty* by Ho Peng Yoke, the following event is described:

> On a Chi-Chhou day in the fifth month of the year of Chi-Ho's reign, 'a guest star' appeared at the south east of Thien-Kaun, measuring several centimetres. After more than a year it faded away . . .

What could this event be? How did it come to be noticed? What was meant by a guest star?

For answers we have to go back a millennium, to the then prevalent Chinese tradition in which the ruling emperor looked to the sky for any 'warnings' from the Almighty, just in case he had happened to stray from the straight and narrow path of fairness and justice. Lest he had to pay a heavy penalty for inadvertently missing such a warning, the emperor made sure that a careful watch was kept on the heavens. It was the duty of the court astrologer to maintain a vigil and inform the emperor of anything unusual. It was in that context that the above event was noticed and duly recorded as stated above. And 4 July 1054 is the date in our modern calendar corresponding to the Chinese record. The word 'guest star' indicates that the star did not exist in the sky prior to the event; more correctly, it had not been observable before. Similarly, after the event was over, the star disappeared from the heavens. The Chinese had the custom of describing such transient objects as guests in the sky. The sighting of this object was also recorded in Japan, where too the astrologers kept fairly meticulous records of the heavens.

Indeed, the star, which perhaps had been previously too faint to be seen, became so bright initially that it could be seen even in daylight, while at night it was five times as bright as the planet Venus in the early morning or late evening. Indeed, when it was at its brightest, one could read by its light at night.

The guest star, however, did not maintain its initial brightness and began to fade. With the help of the old records again, we can deduce today that the object was visible in the daytime for about twenty-three days and was visible at night for about six months. Eventually, within two years, it ceased to be visible. The recorded direction of the object implies a location at Zeta Tauri in the constellation of the Bull. What do we see there today?

Figure 3.1 shows the photograph of that site where, of course, with the naked eye we do not see anything. The photograph shows a remarkable cloud-like structure with several filaments sticking out. Because its shape reminded astronomers of a crab, the object was given the name the Crab Nebula. Certainly, whatever is going on there now must be still pretty violent, judging by its highly disturbed appearance.

We will return to this remarkable picture later. We first look at another bit of evidence of its observation, coming from an altogether different part of the world.

PICTURES ON ROCK

In 1955, William C. Miller published a leaflet under the auspices of the Astronomical Society of the Pacific, presenting evidence that the

Figure 3.1: The Crab Nebula is the expanding remains of a star that was seen to explode by Chinese astronomers in the year 1054 AD. The picture has been reconstituted by David Malin from one taken from the Hale Telescope in the 1960s (courtesy of David Malin, Jay Pasachoff and the California Institute of Technology).

Pueblo Indians in North America had witnessed the event of 1054 and recorded it not on paper but through pictures on rock, pictures which have survived until today.

In Figures 3.2 and 3.3 we see two different types of picture. In the first figure we see a *pictograph*, which is an image made on rock with paint or chalk (or, with a rock that writes like a chalk). This is found in the Navajo Canyon area. The second figure shows a *petroglyph*, that is an image chiselled on rock with a sharp implement. It is from the White Mesa region. The crescent is the Moon, of course; but what is the round object near it? Also, why are the crescents facing the opposite ways in the two pictures?

From old Chinese records one can easily check that the Moon was in a crescent shape when the object was first seen and was at its brightest.

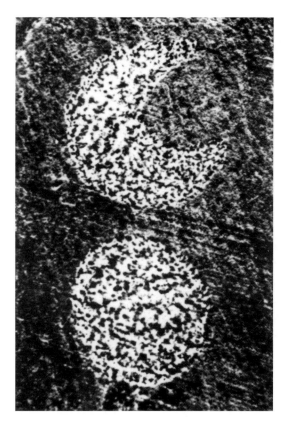

Figure 3.2: The pictograph from the Navajo Canyan, which may record a unique astronomical event seen by the Pueblo Indians in the year 1054 AD (photograph by the late William C. Miller).

Figure 3.3: A petroglyph recording the same event as depicted in Figure 3.2, found in White Mesa (photograph by the late William C. Miller).

The guest star could have been near enough to the Moon for its identi-fication with the round object in these pictures. More- over, these pictures were found in places from where the Eastern horizon was clearly visible. Bearing in mind that such a sight would have been seen near the Eastern horizon, one can attach significance to the loca-tions of these pictures.

Could these pictures represent another more common sight known to observers, namely the occultation of Venus? Miller thinks not, because such occultations occur once in a few years and one would therefore have expected many more such pictures in the area. Rather, one may conclude that the tribes were not routinely interested in astronomy, but were sufficiently impressed by this particularly rare event to have immortalized it on rock.

As to the opposite orientation of the crescent, Miller feels that the artists may have drawn one figure by looking at the original over their shoulders and may have experienced a left–right ambiguity. How would you draw a crescent Moon with your back to it, looking over your shoulder? Try it!

SIGHTING IN THE MIDDLE EAST

On 29 June 1978, in a letter to the prestigious journal *Nature*, Kenneth Brecher from the Massachusetts Institute of Technology and Elinor and Alfred Lieber from Jerusalem presented evidence that the same remark-able sight was seen and recorded in the Middle East by a Christian physician from Baghdad, named Ibn Butan. Although not a professional astronomer or astrologer, Ibn Butan, like his contemporary physicians, was interested in the possibility that diseases on the Earth could be related to cosmic events. Ibn Butan's biography was recorded in a bio-graphical encyclopaedia prepared by Ibn Abi Usaybia around 1242 AD, in which his report is reproduced. Some extracts out of that report are illuminating:

> One of the well-known epidemics of our own time is that which
> occurred when the spectacular star appeared in Gemini in the year
> 446H. In the autumn of that year fourteen thousand people were buried
> in the Church of Luke, after all the cemeteries in Constantinople had
> been filled . . .. As this spectacular star appeared in the sign of Gemini

. . . it caused the epidemic to break out in Fustat when the Nile was low, at the time of its appearance in the year 445H . . .

Here the year is measured on the Islamic Hizri calendar, according to which the year 446 corresponds to the period 12 April 1054 to 1 April 1055 AD, which encompasses the dates when the Chinese saw the guest star. The apparent discrepancy with the year 445H relating to the Nile valley was explained by the authors as due to a copying error on the part of Ibn Abi Usaybia. For, elsewhere in the same encyclopaedia the date checks out to be 446H. Ibn Butan seems to imply that this event occurred in the summer and caused the epidemic in the following autumn, when the Nile was low. This places the event in the summer of 1054, which agrees with the more precise Chinese date of 4 July 1054 AD.

One additional point needs to be sorted out. The Crab Nebula is in the constellation Taurus, whereas Ibn Butan refers to Gemini. However, if one takes into account the steady precession of the Earth's rotation axis, the Crab Nebula would have appeared in Gemini about a thousand years ago.

We thus have three different sources of information about the sighting of a unique cosmic event, from China and Japan in East Asia, from the Middle East in West Asia as well as from the American continent in the western hemisphere. Why are there no records from India or Europe? In India astronomy was flourishing and such an event would have been witnessed at least in some part of the subcontinent, despite the fact that July falls in the monsoon season. The explanation may be that there is not much written tradition in India dating from that time, the emphasis of scholarship then being on reading the ancient texts rather than creating new ones. Nevertheless, some attempts are being made to trace old records of the period which might contain at least oblique references to the event.

And Europe? Why, with their long tradition of keeping and writing manuscripts, did Europeans fail to record this event? Here astrophysicist Fred Hoyle and historian of science George Sarton have independently argued that the religious beliefs of those days assumed that God had created the cosmos in perfection and as such no new phenomena, such as this one, would be considered credible enough to be documented. So the scholars in the monastries chose to ignore what they saw. Perhaps!

But let us now turn to the modern interpretation of this event.

THE CRAB SUPERNOVA

Around 1731, an English physician and amateur astronomer named John Bevis found a bright nebula in the Taurus constellation. In 1758, Charles Messier began his famous catalogue of bright nebulous objects in the sky with this bright object labelled M1. Figure 3.1 shows this remarkable object. As mentioned above, in the middle of the last century it acquired the name *Crab Nebula* because of its crab-like filamentary structure. With its direction matching that of the ancient Chinese records, and its physical environment consistent with the remnant of what that event was, astronomers are sure that the guest star did not actually go away, but is still around in the shape of the Crab Nebula. The approximate distance of this nebula from us is 5000 light years, while the extent of the whole structure of Figure 3.1 is as large as 5 to 10 light years.

So this is what remains today of the event that was witnessed by the Chinese nine and a half centuries ago. Before analysing the event itself, let us make a short diversion which may bring out an element of caution that needs to be exercised by the astronomer when interpreting cosmic photographs.[1]

Misleading photographs

Figure 3.4 shows a photograph of a woman side by side with a young girl. The normal interpretation of a photo of this kind would be that the woman is the mother of the little girl. But what if I tell you that it is the other way round? Impossible, you would say, . . . unless the photographs were taken at different times and put together. The photograph of the mother was taken when she was a little girl and that of the daughter taken recently.

Astronomical photographs are often of this kind. When the image of a star or a galaxy appears on a photographic plate, it is imprinted by the light that has reached the plate from the source object. If the object is located, say, a thousand light years away, this light will have taken a thousand years to make the journey. In other words, the photograph tells us what the source looked like *a thousand years*

---

[1]We remind the reader that a light year is the distance travelled by light in one year; it approximates to about ten thousand billion kilometres.

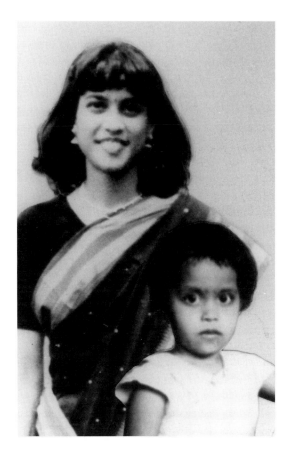

Figure 3.4: In this mother–daughter photograph, who is the mother?

ago, and not what it looks like *now*. So if we see two stars together on the plate, we are not seeing them as they are today, but as they were when their respective light left them to reach us today. Thus a nearby star may look older than a distant star but reality may be the other way round.

Returning now to the Crab Nebula, the object we see in the photograph is located, some 5000 light years away. So when the Chinese saw the 'guest star' in the year 1054 AD, the event had taken place 5000 years previously. Similarly, when we look at Figure 3.1 today we see what was going on in the source 5000 years before today. If we wish to know what is going on there *now*, we will have to wait for another 5000 years.

Exploding stars

Having cleared up the question of this time factor, let us now find out what actually happened when the Chinese saw a star appear and fade away. Piecing together all the written records of the event and relating them to modern theories of stars, the answer is that a star became a *supernova*, throwing out the bulk of its outer envelope in a gigantic explosion.

Why did the star explode? Was it an exceptional event, or do all stars explode? And have similar explosions been seen by astronomers in later years?

We will take up all these questions, although not necessarily in the order they are posed. For example, taking up the last question first, two other similar events have been seen subsequently in our Milky Way Galaxy. In the year 1574 AD the famous astronomer Tycho Brahe observed a supernova, and three decades later, in 1604 AD, Tycho's erstwhile assistant and a distinguished astronomer in his own right, Johannes Kepler, observed another. No supernova in our Galaxy has been observed since then, and in fact since the use of the telescope for astronomy (in 1609 AD). This does not mean that supernovae in our Galaxy occur with a frequency of once in a few centuries. They are believed to be much more common: on average, one star in the Galaxy explodes about every twenty years. As explained in Figure 3.5, because the galaxy is vast and light from other regions gets absorbed most of these events are blocked out from our vision. The three supernovae we could see were in the part of the galaxy readily accessible to us.

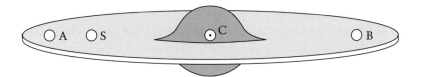

Figure 3.5: Our Milky Way Galaxy is a collection of some one to two hundred billion stars spread over a disc with a slight bulge at the centre. We are located about two-thirds of the way from the centre towards the edge of the disc, shown in the above figure by the letter S for the Sun. The distance of S from the Galactic centre C is more than 30 000 light years. A typical supernova may occur near the central part of the Galaxy or at a point farther away in the disc such as B. Because of absorption by the intervening matter in the Galaxy, such a supernova may not be visible from S. However, supernovae near us, at points such as A, will be visible, but their numbers will be rather small.

Nevertheless, supernovae in other galaxies have been observed every year and are labelled each year chronologically by the letters of the alphabet. Thus Supernova 1987A was the first supernova observed in the year 1987. We will have something more to say about this particular supernova later on.

We now come to the question of *why a star explodes*.

THE EVOLUTION OF A GIANT STAR

In his book published in 1606, called *De Stella Nova* (meaning, *On the New Star*), Kepler conjectured that a supernova may be the result of some random concentration of particles of matter in the heavens. Presenting what he said was

> . . . not my own opinion, but my wife's: Yesterday, when weary with writing, I was called to supper, and a salad I had asked for was set before me. 'It seems then', I said, 'if pewter dishes, leaves of lettuce, grains of salt, drops of water, vinegar, oil, and slices of egg had been flying about in the air for all eternity, it might at last happen by chance that there would come a salad.' 'Yes', responded my lovely, 'but not as nice as this one of mine'.

In the modern version, a supernova arises as the end state of the evolution of a very massive star, a state which is reached by a red giant star when it is no longer able to maintain its equilibrium. How does this state come about?

In the last chapter we discussed the red-giant state of a star such as the Sun, reached after the star has finished its hydrogen fuel and has switched on to another, namely, the fusion of helium. We found that when this change comes about in the star's interior, its outer envelope swells. The expansion of the gases in the envelope lowers its surface temperature, with the result that the star now looks bigger but redder.

We take up the story from this point. Figure 3.6 shows the state of the star after it has consumed all the core helium in this fusion process. Its central part now contains carbon surrounded by a shell of helium that is not hot enough to be fused, which in turn is surrounded by an envelope of even cooler hydrogen. Because it cannot any longer draw upon its helium reservoir, the star once again finds itself at a crossroads.

Recall that it was the central energy-generation process that maintained the high temperature and pressure of the core and kept it in

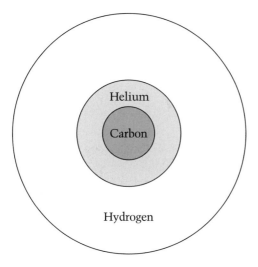

Figure 3.6: The three-layered structure of a giant star which has carbon at the centre, helium in an inner shell and hydrogen outside.

equilibrium against the inward tendency to contract under its own force of gravitation. With the energy source switched off, there is nothing to prevent the core's inward contraction. And as it does so, again a new development intervenes.

Because of its contraction, the core gets hotter and reaches a level when a new fusion reaction begins to operate. This reaction draws upon the carbon nuclei in the core and the helium nuclei that are also around, to make a yet bigger nucleus, that of oxygen (see Figure 3.7).

The effect of this reaction is threefold. First, of course, by supplying a new source of energy it enables the star to shine with renewed vigour and increased luminosity. Second, it stabilizes the core, that is, it puts brakes on its contraction by providing adequate pressures. Third, it

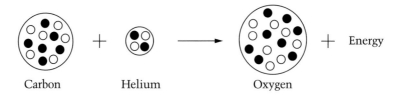

Figure 3.7: The carbon nucleus has 12 particles while the helium nucleus has four. The result of the fusion is to produce an oxygen nucleus with 16 particles. The protons are shown by filled circles and neutrons by open circles. This process releases energy which enables the star to continue shining.

makes the envelope expand further in order to adjust to the new internal pressures. And because of its expansion the envelope cools and looks even redder. As shown in Figure 3.8, the star moves on the H–R diagram further up and towards the right.

Let us pause and comment on a rather peculiar behaviour of the star, when judged by the standards of our daily experience. Our experience tells us that when we put a hot body in contact with a cool body, heat passes from the former to the latter, with the result that the hot body gets colder and the cool body gets warmer, until they both have the same temperature.

Now imagine a thought experiment, in which we connect a hot star to a cool star by a conducting wire. We expect that heat will pass from the hot star to the cool one, and it indeed does. But when the hot star loses energy in this way, it finds that its internal pressures have diminished and therefore its gravitational force pushes it inwards until a new state of equilibrium is reached. In this state, the star has become hotter again, because of its compression.

Likewise, the cool star gains energy, which boosts its internal pressures and makes it expand to a new state of equilibrium. In this state, because of expansion, the star is cooler than before. In other words, the hot star has got hotter and the cool star has become cooler!

Although we cannot achieve the exact conditions of this thought experiment in real life, we come close to it in the red giants. Notice that the core and envelope of the star are in contact and while the core gets hotter at each new stage, the envelope gets cooler.

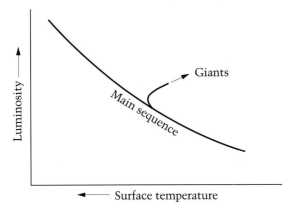

Figure 3.8: This H–R diagram shows how the star moves along the giant branch in the direction of the arrow.

This strange behaviour is, of course, brought about by the force of gravity, which always dictates the equilibrium condition of the star. We shall discuss even stranger effects of gravitation in Chapter 5.

THE ORIGIN OF CHEMICAL ELEMENTS

Returning to the evolving star, we will sooner or later encounter the question again: What happens when the carbon fuel is exhausted? This is bound to happen eventually. When it does, the outcome is again predictable. The core contracts and heats up to another temperature high enough to trigger off another reaction. This time oxygen combines with helium to form neon, whose atom has 20 particles in its nucleus. Again the fusion releases further energy, which keeps the star going for another period. And the star advances further along the giant branch in the H–R diagram.

Thus we have a sequence of reactions which build up heavier and heavier nuclei, the number of particles in each successive nucleus being four more than in the previous nucleus, because each time we add four particles through fusion with a helium nucleus. The sequence of elements so created reads like this: carbon (12), oxygen (16), neon (20), magnesium (24), silicon (28), sulphur (32) etc. They form an 'alpha-particle ladder', so-called because the helium nucleus is also known as an *alpha particle*.

How long does this sequence continue? The answer lies in nuclear physics. Let us look at the force which holds together a nucleus.

This force, as we saw in the previous chapter, is a strong attractive force but with a very short range, typically a million billionth part of a metre. The force within this range is stronger than the electrical repulsion that operates between any two protons. So when we start building bigger and bigger nuclei, to begin with it helps to add more and more neutrons and protons, since the nuclear force of attraction not only encourages the addition of more particles to the fold but it also grows in strength.

The work put in by the nucleus in pulling in more particles adds to its store of energy which becomes available to the star to radiate. That is why the fusion of more particles into an existing nucleus keeps the star shining. However, this cannot go on for ever. Just as a big empire begins to lose its cohesion as it stretches out too much, or a fighting army begins to lose its effectiveness if its supply line becomes too long, so does an atomic nucleus begin to destabilize as it grows too big. There are two reasons for this. First, the range of the attractive force between particles is very limited and if two particles are too far apart, they will cease to attract each other. Second, the addition of further protons to the system increases the electrostatic repulsion, which weakens the binding of the nucleus.

Therefore, by the time the number of particles reaches 56, the nucleus has reached a stage where any further addition will be counter-productive. That is, the new nucleus will not hold together as strongly as the previous one and the star will no longer lower its energy by proceeding further along the fusion track. Figure 3.9 illustrates how the binding of the star changes through the addition of more and more nuclei. It rises and then falls.

The nuclei which are at the peak in this binding property are those of iron, cobalt and nickel. Here the star will have reached the end of the road as far as energy production goes. By this time, its core has heated up to a temperature of several *billion* degrees. But there is no further energy source left in it to keep it that way. What happens next?

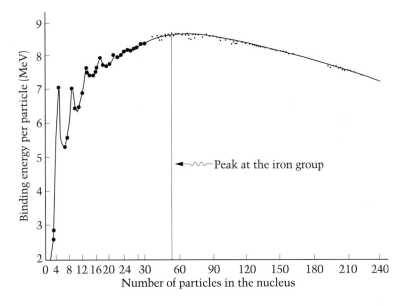

Figure 3.9: This curve shows that the peak of the binding power in an atomic nucleus is attained when the nucleus belongs to the iron group, with about 56 particles in the nucleus.

This question was discussed by four astrophysicists in 1956, within the broader issue of the origin of chemical elements. They were Geoffrey and Margaret Burbidge, William Fowler and Fred Hoyle, and the question they asked was: How did the universe come to have all the variety of chemical elements that we find in it? And can we understand their relative abundances?

For, by astronomical observations one can get a reasonably good estimate of these relative abundances. The means of doing so are to be found in spectroscopy, as we discovered in the context of stars (see Chapter 2). And Burbidge, Burbidge, Fowler and Hoyle (these authors came to be known collectively as B^2FH: see them together in Figure 2.36) worked out the ladder-like process of building up bigger and bigger nuclei up to iron. They also showed that fast and slow processes involving additions of neutrons and their decays could lead to the build-up of heavier elements like gold, silver, uranium, etc. although these processes do not supply any energy to the star.

An anthropic consideration

In the previous chapter we described how Fred Hoyle predicted the existence of an excited level of the carbon nucleus while considering the state of the star that has just exhausted its hydrogen fuel through the fusion process. The reason why such a state had to exist, according to Hoyle, was that it was only then that a resonant fusion of three helium nuclei to that carbon nucleus could take place. The 'resonance' helps in accelerating an otherwise slow process, for the possibility of three helium nuclei getting together at the same time would be relatively rare. And because of such a reaction the star can keep on shining and go into the giant state. The fact that giant stars exist must mean there is a process like this to supply them with energy.

Hoyle had an even stronger motivation for making this conjecture: without it, there seemed no way that elements like carbon and oxygen could be made. Imagine a universe without these elements. A major drawback would be that it would have no life as we know it, for carbon and oxygen are essential elements that go to make up the kind of life seen on Earth. Thus the fact that we human beings are around to observe the universe makes it imperative that the route to making carbon and oxygen must be open!

So, by the time the iron group of elements are made, the star has the 'onion-skin' type of structure shown in Figure 3.10, with elements from the iron group in the central core and progressively lighter elements in the outer shells. The star has reached a critical stage in its existence, for now new factors enter into consideration, factors which determine whether the star will survive or explode.

An analogy with us humans will help. As we get to middle age, our doctors recommend that we keep our weight within a relatively modest limit. Being grossly overweight may invite trouble in the form of hyptertension, heart disease, etc. So the prudent ones shed excess weight

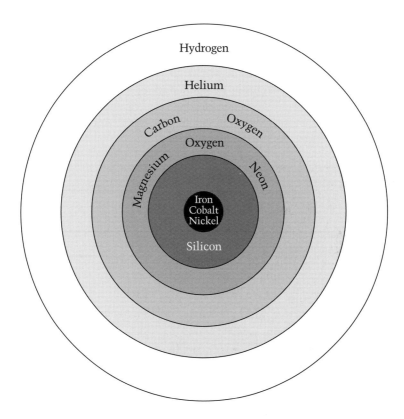

Figure 3.10: The star has a multilayered structure like an onion when it has reached the end of the synthesis of nuclei through fusion. Progressively lighter nuclei occupy successive outer shells.

through exercise and dieting and are perhaps more likely to live long and healthy lives. Those who do not may have to pay the price, an early demise.

A mass limit also exists for stars and it is around six times the solar mass. For stars below this limit during the red-giant phase, there exists a long and relatively unadventurous but safe future. These stars gradually eject small parts of their outer envelope, like smoke rings blown out by a smoker. Figure 3.11 shows such a ring, commonly called a *planetary nebula*: 'nebula', because it has a cloud-like structure; 'planetary', because it is lit up by the parent star just like a planet.

By blowing such 'smoke rings', the star manages to reduce its mass. If it reduces to a low enough mass, it can survive for a long while as a white dwarf. This was the state discussed in the previous chapter, where we found that the critical mass limit for a white dwarf is about 40 per cent above the mass of the Sun and is known as the *Chandrasekhar limit*. The star could also end up in another compact form known as a neutron star, which can be up to twice as massive as the Sun. We shall encounter neutron stars in some detail in the next chapter.

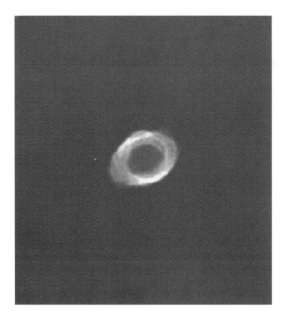

Figure 3.11: The Ring Nebula (CCD image by George Jacoby, NOAO).

Let us now turn our attention to stars which were unwise enough to exceed the critical mass limit when they were giants, a mass limit of about six solar masses.

A very traumatic future lies ahead of them.

The triggering of a supernova

As in the earlier stages when the exhaustion of one type of nuclear fuel led to the contraction of the innermost core, now again the stellar core contracts. But in the earlier situation the resulting higher temperature of the core initiated a new fusion reaction. For red giants whose mass exceeds the Chandrasekhar limit, that possibility is no longer available. As we saw just now, it is not possible to extract more energy by fusion beyond the iron group of elements. Instead, as the core contracts, the iron group of elements break back into helium nuclei as well as free protons and neutrons, with a resultant *loss* of energy in the core. Instead of restoring equilibrium, this process accelerates the core contraction process.

The rapid contraction is often called the *collapse* of the core. And it has serious repercussions for the envelope as well. As the core collapses, the effect of degeneracy pressure akin to that we encountered earlier for white dwarfs (see Chapter 2) comes into play, although in a transient fashion.

In the case of white dwarfs the degeneracy arises because the electrons are packed too tightly together. The laws of quantum mechanics place a restriction on how many electrons up to any specified level of energy can be packed close together in any given volume. Here, for the supernova core, the degeneracy arises from close packing of neutrons. But where do these neutrons come from?

In the core, the break-up of the iron group of nuclei produces free neutrons and protons. In a terrestrial laboratory a neutron does not last very long. Within a few minutes it decays, producing an electron, a proton and a particle called an antineutrino.[2] Therefore, under terrestrial conditions the neutron is not a stable particle. It remains stable, however, within the nucleus of an atom because of the strong force

[2]The neutrino is a particle of matter which is believed to have no mass at rest. It is in fact supposed never to come to rest but always move with the speed of light. Particle physicists, however, do not rule out the possibility that the neutrino may have a tiny mass and it may therefore slow down and have a position of rest also.

operating there. Now as the core collapses, the reverse reaction to neutron decay takes place. The core contains plasma at high density, that is, a mixture of electrons and ions (see Chapter 2), and this mixture also contains free protons. So in the reverse reaction the electron and the proton combine to form a neutron. This reaction also releases a neutrino.

All this happens when the core is collapsing. First, neutrons are formed in the above manner and as their density grows rapidly they begin to generate a strong degeneracy pressure. This pressure puts up a strong resistance to the core collapse. It succeeds not only in halting the collapse but in making the core bounce, much like a ball bouncing off a hard surface.

This takes hardly a few seconds, and the core now starts moving outwards rapidly. The envelope, meanwhile has not had time to react to this rapid development and it experiences a buffeting from the outward moving core (see Figures 3.12 and 3.13). In the language of the physicist, we say that a shock wave is released by this process.

A shock wave is nothing other than a moving surface of disturbance, across which there is a tremendous difference of pressure. In normal physical processes the pressure changes smoothly across a medium whereas in an explosive process the pressure drops across a surface rather sharply. This discontinuous change drives the surface in the direction of the low pressure with great force. This is the shock wave which is released in any explosive process.

In the star, therefore, the shock wave produces a highly disruptive impact on the envelope, blowing it outwards rapidly. *This is the stage when the star is said to explode, when it becomes a supernova.*

Before we discuss the highly picturesque aftermath of the explosion, we should not forget an event that comes as a *precursor* a few moments before the explosion. We refer to the neutrinos that are created when the core matter suddenly turns into a large number of neutrons.

The neutrinos travel out of the star with the speed of light. They have the property that they come out practically unscathed through the entire star, because they interact only very weakly with any form of matter. In other words, the matter en route does not place any obstacle

But this hypothesis has not yet been experimentally confirmed. So we will assume here that neutrinos are always travelling with the speed of light. The antineutrino is a similar particle but made of antimatter. Matter and antimatter annihilate each other producing radiation; thus a neutrino and antineutrino pair will annihilate when brought together.

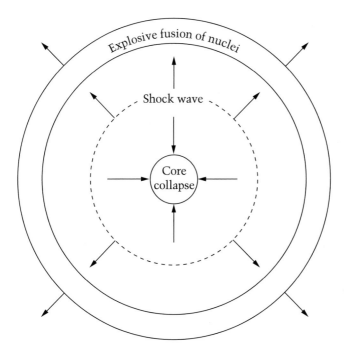

Figure 3.12: This shows how the shock wave generated in the inner regions of the about-to-explode star travels outwards heating the outer layers of the star and triggering bursts of nuclear fusion there.

in their way, as it would if they were escaping particles of light, *photons*. So we have the remarkable result that *just before the star explodes it sends out a large flux of neutrinos, produced in its core*.

We will have occasion to recall this result later in this chapter.

THE AFTERMATH

The shock wave generated at the interface of the core and envelope of the star disrupts the latter and ejects most of it into interstellar space. But before this happens, for a period lasting not more than a few tens of seconds, the outgoing shock wave heats up the outer parts of the envelope.

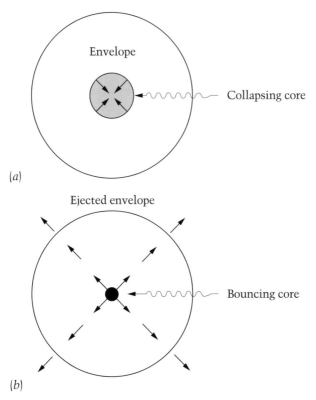

Figure 3.13: The core is moving inwards rapidly in (a), while in (b) it is moving rapidly outwards. The rapidly changing physical scenario on the core–envelope boundary releases a shock wave that ejects the envelope outwards.

Since the star, prior to this catastrophe, had acquired an onion skin structure (see Figure 3.10) with layers of lighter and lighter nuclei farther and farther from the central core, these layers get heated to such temperatures that their nuclei undergo a fusion process. Figure 3.12 illustrates this phenomenon, called *explosive nucleosynthesis*, as it occurs like a burst for a short period of time. Yet it can have interesting implications for the environment of the supernova, as we shall see later in this chapter.

The explosion of the star itself, which dislodges the envelope and ejects it into space, is of course much more energetic than the explosive nucleosynthesis. The energy is in the form of radiation and particles such as electrons, protons, neutrons and atomic nuclei. For a brief moment of glory before death, the star generates so much energy that it outshines the entire galaxy it is situated in, a galaxy that may have more than *a hundred billion stars*. No wonder the Chinese saw the guest star as visible in daylight.

Theorists can calculate how the brightness of the star sharply rises and then falls steadily. Figure 3.14 shows a typical light curve for a supernova. Notice that it rises and falls sharply over a few days, but then there is a steady decline that takes more than a year. Thus for naked-eye observers the star would no longer be visible. The 'guest' will have departed.

The spoon in your hand

Let us pause and think about an everyday matter, about the metal spoon which you use for eating. Where does its material come from?

Well, a stainless steel spoon will have been made in some factory which obtained the basic ingredient, steel, from an iron and steel plant. The plant will have processed the steel from ore dug out of a mine. The mine represents the iron deposit in the Earth. So one might be tempted to answer the question by pointing to the Earth as source.

But that is not the ultimate answer. We ask, how did the iron get into the Earth? The answer may be that it was present in the material in interstellar space out of which the Earth was formed. But, then, how did it get into the interstellar medium?

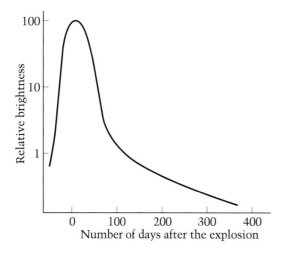

Figure 3.14: A supernova light curve, which shows how the supernova brightens dramatically and dims rapidly, in the initial few days, and then gradually fades over a year or two.

This is where supernovas come in. A supernova, when it explodes, dumps the iron made within it into the surrounding interstellar space. The iron was made in the progenitor star at a temperature of several billion degrees.

So, next time you stir your tea think of the high temperatures through which the material of your spoon has been!

Cosmic rays

The particles and nuclei ejected by a supernova emerge from it with very high energy, so that most of them travel with speeds very close to the speed of light. Where do they go? Once they get out of the hot and turbulent environment of the exploding star, they can travel all over its galaxy. However, they do get deflected by any magnetic field in the galaxy. Thus if we receive a flux of such particles, we cannot be sure that their source lies in the direction in our Galaxy from which they actually approach the Earth.

We do get bombarded by such particles from all directions. They are called *cosmic rays*. The first discovery of cosmic rays was made at the turn of the century. Physicists noticed that their electroscopes, devices which store electric charge, tended to discharge themselves even under the protection of thick lead shields. This was only possible if the discharging was done by a bombardment of fast-moving oppositely charged particles from outside, particles energetic enough to penetrate a lead shield. The physicist C.T.R. Wilson conjectured that these particles might be coming from outside the Earth, although the majority of physicists believed that they came from the Earth's crystal rocks.

If the majority opinions were right, then as one went away from the Earth's surface the intensity of flow of these particles should diminish. In 1910, the Swiss physicist Albert Gockel ascended in a balloon to a height of about 4000 metres and found that the intensity was undiminished. In 1912 Victor Franz Hess went higher, to a height of about 5000 metres, and found that the intensity in fact *increased*. The increase of intensity with height was confirmed to greater heights in subsequent years and it became clear that Wilson's conjecture was correct; the name 'cosmic rays' was coined. These rays contain particles such as electrons, protons, neutrons and atomic nuclei. They may also contain small quantities of antimatter.

Where do these fast-moving particles, which bombard the Earth, come from? As we have seen, in supernovae we have a likely source.

In the post-explosion scenario, the ejecta from the star travel in all directions and some of them may come our way. One can also wonder what would happen if we were located close enough to a supernova for us to receive a large flux of cosmic rays.

The scenario is not very pleasant. For if the flux of cosmic rays were very high, the protective layer of ozone in the atmosphere that surrounds the Earth would be destroyed by the incoming particles. This in turn would cause the ultraviolet radiation from the Sun to reach the Earth in quantities large enough to destroy life on Earth. How close does the supernova have to be in order for this grim possibility to be realized? Closer than about 30 light years, for a supernova as powerful as that in the Crab Nebula. The Crab Nebula is fortunately about 200 times farther away than this limit! Also, within this critical distance, there are not many giant stars that could become supernovae, but who knows . . .?

Imagine, however, that a supernova did go off at 30 light years from us. The light takes 30 years to reach us. So when we see the event here, it has already taken place 30 years earlier. What about the cosmic ray flux? The cosmic ray particles are travelling at nearly the speed of light, but may not travel directly to us: a galactic magnetic field may delay their arrival at the Earth by a few years. The inhabitants of the Earth will have to find some counter-measure against the ensuing catastrophe within this 'grace' period.

SUPERNOVA 1987A

Although supernovae seen in our own galaxy have been relatively rare, we do regularly see supernovae go off in other galaxies. As mentioned earlier, the supernovae seen every year are chronologically labelled in alphabetical order. Let us look at some details of a spectacular supernova seen in 1987. Being the first one of the year, it was catalogued under the name 1987A. The circumstances leading to its discovery were as follows.

On 24 February 1987, Ian Shelton from the University of Toronto, who was a resident astronomer at the Las Campanas Observatory in Chile, happened to notice a newly brightened star in the direction of the Large Magellanic Cloud (LMC). He took a photograph of the star, and this was the first documentation of a new supernova which kept astronomers, the world over, busy with further studies of the remarkable object. For, when compared with the image of the star the previous day, 23 February 1987, Shelton's image was enormously brighter. As was subsequently calculated, the brightness attained by the supernova was about 5 per cent of the *total light of all stars in the LMC put together*!

The LMC is one of two 'clouds' seen by the sixteenth-century explorer Ferdinand Magellan in a voyage which took him to the southern hemisphere. The LMC and the Small Magellanic Cloud are in fact tiny irregular galaxies considered to be satellites to our own Milky Way.

Although Shelton's sighting of the supernova was the first 'news' of a supernova going off in the LMC, it was not the first message of the phenomenon to reach the Earth. We will come back to this cryptic remark shortly.

This supernova turned out, in many ways, to be a fertile ground for testing astrophysical theories.

The star which exploded, Sanduleak (catalogue name Sk-69 202), was a blue supergiant star with a surface temperature of 20 000 K and luminosity 40 000 times that of the Sun (see Figure 3.15). Its radius was estimated at 15 times the solar radius and its mass at the time of formation 17 times the solar mass.

Figure 3.15: The star Sanduleak shown on the right exploded as a supernova seen on the left (Anglo-Australian Observatory: photograph by David Malin).

These details could be estimated owing to the fortunate circumstance that this supernova was practically on the doorstep of our Galaxy. It was at the relatively modest distance of about 170 000 light years and was visible relatively easily.

Astrophysicists estimate that the collapse of the core which triggered off the explosion occurred a few hours before the explosion. If it were possible we would have witnessed that event at 07.35 Universal Time[3] on 23 February 1987. Although, we cannot 'see' inside a star, there is another way the information was in fact brought to us. *A large flux of neutrinos was released at the time of collapse of the core.*

As luck would have it, two laboratories, one in Kamiokande, Japan, the other, known as IMB, in the United States, had neutrino detectors set up. Both detected 10 neutrinos a few hours prior to the visual sighting of the explosion. This was exactly as expected. The significance of this finding was, however, appreciated only later, after the visual sighting of the supernova was announced.

The supernova 1987A was, of course, monitored optically by several observers and the light emission increased rapidly in a day, to a thousand times that of the original star. The radial size also increased from 15 solar radii to the size of the orbit of Mars. This was when it became a supernova. When it was discovered by Shelton optically, 22 hours had elapsed since the core collapse.

The nuclei produced in a supernova include some which decay through radioactivity. The decay products include high-energy gamma rays. Not all the gamma rays escape without energy loss, but some do and these were detected initially by the *Solar Max* satellite and later by balloon experiments. This was an additional check on the theory of supernova explosion.

Between the summer of 1987 and 1988 the total luminosity of the supernova, arising from the gamma rays which lost energy to visible light and infrared, declined. The characteristic period of this decay was about 114 days. The rate at which this decay took place, and other data, provided valuable checks on theories of stellar nucleosynthesis.

Thus the appearance of Supernova 1987A showed how in modern times astronomers, with many different checks, can test and improve their theories.

---

[3]Universal Time is the clock used by astronomers worldwide to record events. It is the Greenwich Mean Time used earlier with a few technical corrections.

IN MY END IS MY BEGINNING!

What we have described here may be termed the death of a star. In Chapter 2 we discussed the current ideas on how a star produces and radiates energy. But one aspect of a star's life has so far gone without a mention. How is a star born? The present understanding of this phenomenon is briefly as follows. We describe it here because, in a strange fashion, the violent death of a star can trigger off the birth of a new generation of stars.

In the vast interstellar space there are huge clouds of gas, which are diffuse and essentially dark. Because of infrared and microwave astronomy, however, the structure of these clouds has become better known. Figure 3.16 shows the Orion Nebula, which can of course be seen through an amateur telescope. The bright parts of the nebula are lit by the stars present in the cloud.

But there is more to it than meets the eye, as shown in the figure. For, infrared astronomy has shown that there are pockets from which strong infrared emission is coming. And microwave or milli-metre wave astronomy has revealed the existence of molecules of carbon monoxide. The discovery of chemical molecules came as a surprise to astronomers in the sixties, for the majority of them believed that interstellar space would only contain simple elements such as hydrogen. But it is the infrared that concerns us here: the theory of star formation tells us that infrared emission comes from newly born stars.

Indeed, stars are believed to form in large molecular clouds, in parts which are denser than their surroundings. The belief is that the denser portions will contract because of their relatively larger inward gravita-tional pull. Such portions form into balls that go on shrinking and get-ting warmer and warmer inside. These are *proto-stars*, which will become real stars when their inner core gets hot enough to trigger off the nuclear fusion reaction. Until then, these warmish stars essentially radiate at infrared wavelengths.

This scenario also envisages the formation of planets along with a star. If the portion of a gas cloud that becomes a star were spinning about an axis, then its equatorial region would spread out and become a large disc surrounding a central bulge, as shown in Figure 3.17. It is believed that the central bulge becomes the star while the planets form after the break-up of the disc. As the disc was spinning about the bulge, so the planets orbit the central star. This idea received a boost in 1983

Figure 3.16: The Orion Nebula showing portions where carbon monoxide molecules are found and where infrared sources exist (courtesy of the observatories of the Carnegie Institution of Washington).

with the discovery by the Infrared Astronomy Satellite (IRAS) of such proto-planetary discs around a few stars (see Figure 3.18).

So, a cloud like the Orion nebula is a gigantic stellar nursery, one amongst many in the Galaxy. Thus the star-formation process goes on

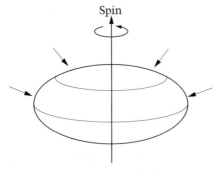

Spin

Contracting cloud

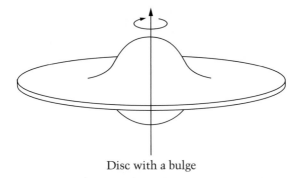

Disc with a bulge

Figure 3.17: The contracting and spinning cloud spreads out in the form of a disc around a central bulge. The bulge goes on to become a star while the disc breaks up into planets.

side by side with the evolution and death of older stars. But the question that troubled astrophysicists was this. Would the force of gravity of a dense portion of a giant molecular cloud begin the shrinking process on its own? The force of gravity in the early stages when the cloud is very distended is not particularly strong.

We are now able to answer this question. *The formation of new stars from interstellar clouds could be assisted or even induced by the explosion of a nearby supernova.* We will describe two lines of evidence that corroborate this idea.

The first piece of evidence was brought by a meteorite that fell in 1969 in the Mexican village of Pueblito de Allende. Known as the Allende meteorite, it showed certain peculiarities in its nuclear composition (see Figure 3.19). It is these peculiarities, known as *isotopic*

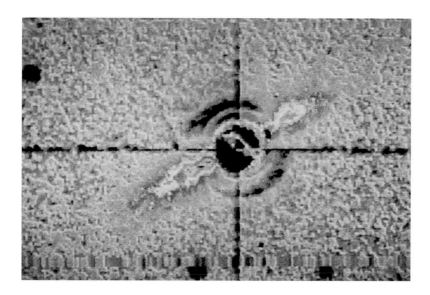

Figure 3.18: The IRAS picture of the disc around the star Beta-Pictoris (courtesy of NASA).

Figure 3.19: The Allende meteorite.

anomalies, that hold important clues to the origin of our solar system.

Different *isotopes* of an element contain nuclei with the same number of protons but with differing numbers of neutrons. For example, the metal aluminium from which our pots and pans are made is the stable element containing in its nucleus 13 protons and 14 neutrons. It is written as ^{27}Al. It has an unstable isotope called ^{26}Al which contains 13 protons and 13 neutrons. Because the chemical properties of an element are determined by the number of *charged* particles in its nucleus, both ^{27}Al and ^{26}Al would have the same chemical properties. But their nuclear properties are different.

The unstable ^{26}Al is a radioactive substance and has a 'half-life' of 720 000 years. That is, if we have in store 100 nuclei of ^{26}Al, on average half of them (50) would decay by radioactivity in this period. The main decay product is a radioactive isotope of another element, magnesium, written as ^{26}Mg. This magnesium nucleus contains 12 protons and 14 neutrons. Thus one of the protons in the original aluminium nucleus has been changed to a neutron. In addition, a positron (e^+) and a neutrino (v) are also released.

Now the striking feature of the Allende meteorite was that it contained certain isotopes in proportions very different from those normally found in the components of the solar system. These differences in abundance are known as *isotopic anomalies*. In the Allende meteorite was found an anomalously high proportion of ^{26}Mg. Why should this happen?

The above question and its answer can be better understood with an analogy. Suppose a country has imposed gold control laws under which its citizens are not permitted to hold pure gold beyond a prescribed quota. If a spot check of a section of the population turns up a person possessing gold way above this quota the question arises: How did that person acquire so much gold? Investigations may eventually lead to the discovery that he had smuggled the gold from another country where it was easily available. So the question astrophysicists asked about the Allende meteorite was: Where and how did it acquire anomalously large stores of magnesium 26? Their investigations, which are described below, were no less exciting than those unearthing the sources of contraband.

There are many processes which could in principle produce the extra ^{26}Mg. However, a clue to the correct answer was revealed when the mineral grains of the meteorite were carefully analysed. It was then found that the abundance of ^{26}Mg was correlated with that of ^{27}Al, thus suggesting some link between magnesium and aluminium. As we have just seen, the link is via ^{26}Al, which decays to ^{26}Mg.

So it was concluded that either the ^{26}Al somehow got into the meteorite material and then decayed there, over a period of 720 000 years or so, or the meteorite was made from a portion of interstellar medium already containing ^{26}Mg formed from the decay of the ^{26}Al present in the medium. The latter scenario seemed more feasible but it implied that the meteorite must have been formed *soon after* the contamination of the interstellar medium with ^{26}Mg, for otherwise the constant churning up of the interstellar medium by cosmic processes would have wiped out the signature of any *old* contamination. Hence the conclusion that the formation of the meteorite would have taken place soon after the deposit and decay of ^{26}Al in the interstellar medium. What cosmic process could have deposited this isotope of aluminium in the interstellar space?

This is where the supernova comes in. Notice first that the ladder of successively bigger nuclei, which we described earlier as forming in successive fusion reactions, increases the number of particles in the nucleus by steps of four. Thus we get ^{12}C, ^{16}O, ^{20}Ne, ^{24}Mg and so on. ^{26}Al does not fit into this sequence. But it can be made during the explosive nucleosynthesis phase of the supernova that we described earlier. In this phase free neutrons (n) and protons (p) can be added to form nuclei off the alpha-particle ladder. For example, ^{26}Al is made from ^{24}Mg by the series of reactions illustrated by Figure 3.20. There are also other ways of making ^{26}Al in this phase of the supernova. These ejecta from the explosion can very well contaminate the nearby interstellar space.

It is suggested that the isotopic anomalies of the Allende meteorite, like that of ^{26}Mg which we discussed above, arose because a supernova went off in the neighbourhood of the gas cloud out of which the solar system formed. And the occurrence of the supernova cannot have been very far back in time before the formation of the solar system. If, for example, the time gap between the supernova explosion and the formation of the solar system were a million years or more, all signature of supernova contamination would have been wiped out.

This evidence from the Allende meteorite therefore links the origin of our solar system with a comparatively recent supernova. It is possible that the presence of the supernova in the vicinity of the solar system was purely accidental, as also the timing of its explosion just before the

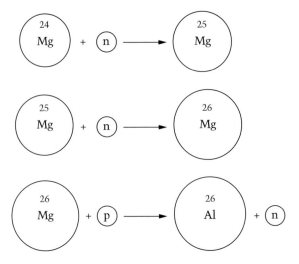

Figure 3.20: Diagram showing how the isotope ^{26}Al of aluminium can be formed from the isotope ^{24}Mg on the alpha particle ladder, by the addition of neutrons and a proton.

solar system began to form. However, supernovae being somewhat rare objects, there may well be more to it than meets the eye. Indeed, there is a physical argument to suggest that the explosion of a supernova triggers off the process of star formation in its vicinity. Let us briefly look at this argument before examining the second piece of evidence.

Recall first that the explosion of the star was caused by a gigantic shock wave that originated in the star's core and travelled outwards. The wave does not terminate at the star's surface but continues moving outwards. As it recedes from the centre of explosion its intensity naturally declines. However, in the immediate vicinity of the star it may still be very powerful. Such a wave impinging on a nearby interstellar cloud can therefore give it a strong push. This push is just what is needed to set off the cloud's compression and it resolves the difficulty we mentioned before, that gravitation is initially too weak to start the compression of a large diffuse cloud. The external pressure from the shock wave tilts the balance of all the forces on the gas in the cloud in favour of contraction. Do we have any evidence that such shock waves existed in the vicinity of newly forming stars? Yes! Such evidence was uncovered in 1977 by two astronomers, William Herbst and George Assousa.

Herbst and Assousa examined the vicinity of the astronomical object called Canis Majoris R-1. Shown in Figure 3.21, this is a supernova remnant like the Crab Nebula of Figure 3.1. As in the Crab, there is evidence of outward motion of gas particles, indicating that there was an explosion. Estimates based on these motions show that the explosion took place about 800 000 years prior to the state now being observed in Canis Majoris R-1. More interestingly, new pre-main-sequence stars have been seen in a region not too far from the supernova remnant. These stars, whose ages are believed to be only about 300 000 years, are probably the youngest stars known to astronomers. They have not yet become fully fledged stars by triggering of the nuclear reactors in their cores.

Evidently these stars formed *after* the explosion. How big was the explosion? If we try to work backwards from the present observations of

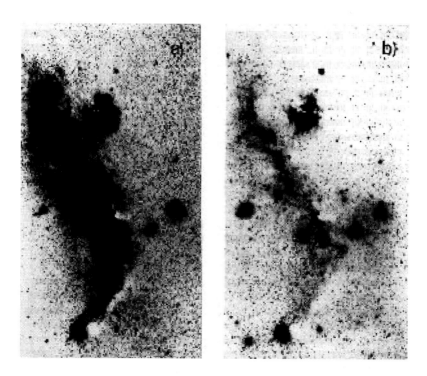

Figure 3.21: Supernova remnant Canis Majoris R-1. In (a) we have the red print, in (b) the blue print. For details see the original paper by W. Herbst and G.E. Assousa in *Astrophysical Journal*, **217**, 475, 1977 from which this photograph is taken.

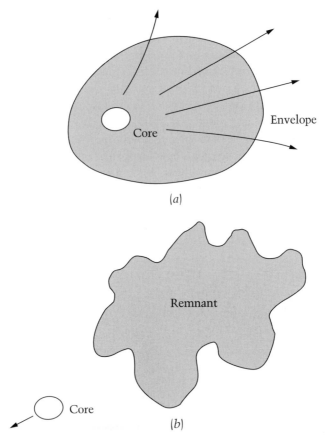

Figure 3.22: A skew explosion of a supernova is shown in (a). It shoots off the envelope in one direction while the core recoils in the opposite direction, like a rifle butt after a shot. This is shown in (b).

outward moving gas, we arrive at a figure for the energy released in the explosion equal to the energy that the Sun radiating with its present power will take around eight billion years to emit! Fantastic though this figure seems in the normal stellar context it is characteristic of the energy in a supernova explosion.

For an exploding star we would also expect evidence in the form of the leftover portion, that is, the inner core. We do indeed see a star, not within the supernova remnant but outside it. This star is found to be moving away from the remnant with unusually high speed. Could this be the star whose envelope was ejected in the supernova explosion? A plausible case for such a course of event can be made on the analogy of a rifle shot. Just as the rifle recoils on firing the shot, the star in question recoiled after ejecting its envelope in the opposite direction. Figure 3.22 shows how a high recoil speed is generated in an oblong asymmetric explosion. The measured speed of the star fits the recoil hypothesis.

So there is suggestive evidence that links the formation of new stars to a recent supernova explosion and it lends further strength to the hypothesis that star formation in general is induced by the explosions of stars of earlier generation. Our story of the star's life has thus come a full circle by linking the destruction of one star with the birth of another!

But it is premature to write off the star at this stage, for there is more to come in its life even after its apparent destruction in a supernova explosion. That story will lead us to another wonder of the cosmos.

Pulsars: the timekeepers of the cosmos (4)

SIGNALS FROM SPACE

Date: 6 August 1967. Location: Cambridge, England.

Jocelyn Bell, a graduate student in the Mullard Radio Astronomy Observatory of the Cavendish Laboratory in the University of Cambridge, was going through transcripts of the data she had collected to monitor the effect of interplanetary scintillation. The record showed a fluctuating signal that could have come from a radio source undergoing scintillation in the anti-solar direction. Shown in Figure 4.1, a pattern of this type coming from such a direction was most unusual.

Scintillation is the phenomenon of the twinkling of a radio source when its radiation passes through a cloud of plasma that is fluctuating. The plasma, a mixture of positively charged ions and negatively charged electrons, is present in interplanetary space. The up-and-down pattern of the source intensity is quite pronounced if the source has a small apparent size, so that it subtends an angle at observers like us of the order of 1 arcsecond. (One arcsecond is a 3600th part of a degree.)

Realizing the potential of this method for measuring the angular sizes of very small radio sources, Antony Hewish at the Mullard had set up an elaborate experiment to survey the sky for scintillating sources; Jocelyn Bell took part in this survey project (see Figures 4.2 and 4.3). When she reported her unexpected finding to Hewish, he realized that the signals needed further checking out for what they were (and for what they were not!).

He therefore initiated an elaborate programme of monitoring this phenomenon to check whether it was due to some electrical interference or a flare star. By 28 November, he and Bell found that what they were looking at was a pulsating source: see in Figure 4.4 a transcript of the first pulsating signals received from the source. And it became

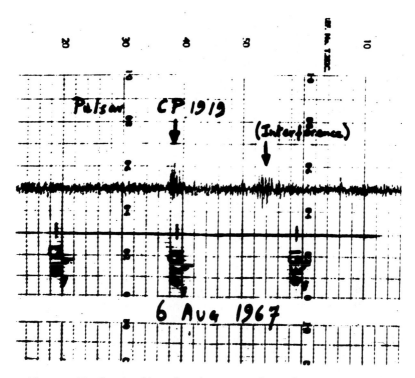

Figure 4.1: The first signal from the pulsar CP 1919 detected on 6 August 1967 by Jocelyn Bell. (Photograph by courtesy of A. Hewish.)

abundantly clear that here was an astronomical phenomenon the like of which had never been seen before.

The results of a preliminary analysis were presented by Hewish on 20 February 1968, at a crowded Cavendish seminar entitled 'Discovery of a new type of radio source'. I recall several of us from the Institute of Theoretical Astronomy, including its founder Fred Hoyle, going over for the lecture. Working out on Madingley Road along the western outskirts of Cambridge, we did not as a rule attend seminars in the old Cavendish laboratories in the centre of the town. But that day was different, as we had an inkling that something of an exceptional nature was going to be announced by the speaker.

There certainly was an air of expectation and one unusual aspect noticed by the assembled audience was that the blackboard of the venerable Maxwell Lecture Theatre carried cutouts of little green

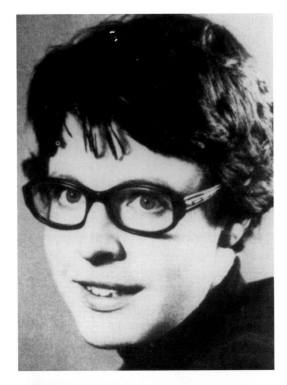

Figure 4.2:
Jocelyn Bell.

Figure 4.3:
Antony Hewish.

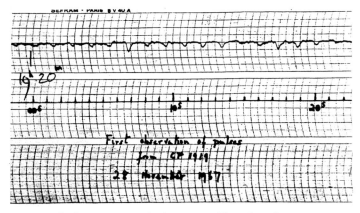

Figure 4.4: The first transcript of pulses received from the pulsar CP 1919 on 28 November 1967. (Photograph by courtesy of A. Hewish.)

men! Were we going to hear about signals from advanced extra-terrestrials?

We did hear about signals, signals first spotted by Jocelyn Bell and then carefully tested for authenticity of their extra-terrestrial origin by her research guide Hewish along with other colleagues including Bell. The signals were in the form of extremely regular radio pulses, as Bell had noticed earlier. Their period, that is the time difference between two successive pulses, was measured to be 1.337 301 1512 second. The fact that one can quote the period to an accuracy of ten decimal places was remarkable and quite unprecedented for any astronomical observation. What was the source of these highly regular radio pulses?

Hewish's conclusion that day may have disappointed the science fiction buffs, for he did not believe that they were sent by some intelligent supercivilization. Why? Because, such a civilization would be located on a planet going round a star, not on the star itself. (A star would be somewhat too hot for habitation by living beings!) In such a situation, the planet would be alternately moving towards us and away from us and this would cause the pulse frequency to rise and fall.[4] No such effect was seen and the frequency held steady. Thus the source had to be an object that did not have such orbiting motion.

[4]This is well-known Doppler effect, first explored in the last century by Christian Doppler for sound waves. This effect causes the pitch of the sound of an approaching source to rise and of a receding one to fall. The same effect for light or radio waves translates into the rise or fall of the frequency.

What could the source be? Since the pulses were of such a short duration, the source had to be very compact. A large extended source is not expected to send out such signals because any coherent physical changes in it would be expected to have a much longer repeat time (period). Therefore this source had to be different from any sources of radio waves known to the astronomers up till then. With regard to compact sources, white dwarfs or neutron stars were considered strong candidates.

That day we came away from the Hewish seminar with the exciting prospect that astronomers had encountered a new challenge. To think of a scenario for a phenomenon having such a high degree of temporal regularity, with such a small period, was not going to be easy.

This remarkable source was christened a 'pulsar' to underscore its pulsating character. It was given the catalogue name CP 1919, the letters standing for Cambridge pulsar and the numbers relating to its position as measured by an astronomical coordinate, the sidereal time (19^h 19^m) in the sky.

Soon after the Cambride discovery was announced, the radio telescopes at Jodrell Bank near Manchester and at Arecibo, Puerto Rico, were used to look for other similar sources and several were found. Up to the present time, more than 600 pulsars have been found. Each is now catalogued with the letters PSR (for pulsating source in radio), followed by two juxtaposed numbers which tell the astronomers its position on the sky.

Let us now see why the pulsar is one of the most exciting objects in our Galaxy, an object that not only has remarkable observational features but which also demands applications of physics at the very frontier of knowledge. In 1974 Hewish was awarded a Nobel Prize for this discovery and he concluded his Nobel Lecture with the words:

> In outlining the physics of neutron stars and my good fortune in stumbling upon them, I hope that I have given some idea of the interest and rewards of extending physics beyond the confines of laboratories. These are good times in which to be an astrophysicist . . ..

THE NEUTRON STAR

Of the two compact candidates for a pulsar, we have already come across white dwarfs in Chapter 2. Thanks to the early work of R.H.

Fowler and S. Chandrasekhar, the nature of white dwarf stars had been explained by the mid-1930s. Although there were doubts cast on the validity of Chandrasekhar's work by no less an expert than Eddington, the concept of the Chandrasekhar limit became well established within a decade or so.

This limit, as we found in Chapter 2, basically tells us that no star with a mass exceeding it can exist as a white dwarf. Compared to the Sun's mass, this limit stands about forty per cent higher. Certainly we do not find any white dwarfs above this limit.

The value of this limit was computed by Chandrasekhar by taking into consideration the behaviour of matter when it is compressed to a very high density, something like a million times that of water. This is the kind of density believed to exist in a white dwarf. Thus one litre of white dwarf matter will contain a mass of a thousand tons! At this density the electrons in the matter become *degenerate*. That is, their number per unit volume becomes so large that some basic rules of quantum theory, imposing restrictions on the close packing of particles, become applicable.

A similar situation should also in principle exist if we had instead a close packing of neutrons. We saw in the last chapter that just before a supernova explosion, the core of the supernova builds up into such a state. *And after the envelope is blown off into interstellar space, the core survives with neutrons as the main ingredient.* The core may oscillate for a while before settling down to a state of equilibrium, when it consists mainly of tightly packed neutrons.

This is how a *neutron star* is born.

Here is a situation similar to that discovered by Chandrasekhar for white dwarfs. There is a limit to the mass of a star that can be supported by degenerate neutrons. The limit is not very clear cut because the physical properties of matter at densities of *millions of billions times the density of water* are not yet well understood. But experts agree that the mass limit is close to twice the mass of the Sun. Only stars with masses below this limit can maintain their equilibrium as *neutron stars*.

Figure 4.5 shows a schematic picture of how a neutron star is composed of different forms of matter ranging from a very dense state in the centre to a comparatively rarefied state in the outermost layers. It should be remembered, however, that even these more rarefied outer layers are as dense as some of the inner layers of a white dwarf! Notice also that the star in Figure 4.5 is 40% more massive than the Sun but has

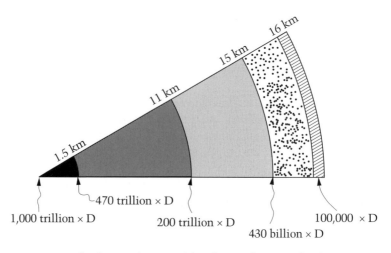

Figure 4.5: A wedge showing the internal distribution of matter and its density in a neutron star. D is the density of water and 1 trillion is a million billion.

an overall radius of *only* 16 kilometres. (The Sun's radius is 700 000 kilometres.)

How would we actually detect a neutron star? As we mentioned before, it would be too faint and too hot at the surface to appear on the standard H–R diagram. Are there other ways of establishing its existence in a given part of the Galaxy?

In 1964, Fred Hoyle, John Wheeler and I suggested in a paper in the scientific journal *Nature* that a neutron star might be detected through its oscillations. As mentioned earlier, the star is formed from the contracting core of a supernova and the core oscillates before it settles down to a static form. Such oscillations of the star could continue for quite a long time since there is considerable dynamical energy to be got rid of. What we argued was that the energy could be dissipated by electromagnetic waves generated in the vicinity of the star through its oscillations. For, it is expected that a very large magnetic field would be present in the vicinity of the star, and this would take part in the oscillations and produce electromagnetic waves. The wavelength of the radio waves emitted, as calculated by us, was very long – about 3000 metres.

We further argued that such long waves would be reflected back by any gas cloud with sufficiently high particle density. But, in the process of reflection, the waves would push the cloud outwards along their original direction prior to reflection. The filaments in the Crab

Nebula (Figure 3.1) appear to be moving outwards from the source probably because of such an effect.

As it turned out, many parts of the above scenario were correct. Thus the assumption of strong magnetic fields near neutron stars, made in the above picture, is now known to be true. A normal star may possess a weak magnetic field. But, as it contracts, the magnetic lines of force passing through the star are squeezed, along with the matter in the star. Normally, densely packed lines of force indicate a strong magnetic force. Thus in the contracting core that becomes the neutron star, the squeezing is quite strong and this produces magnetic fields as high as a thousand billion gauss near the star's surface. (For comparison, the magnetic field near the Sun's surface is only 1 to 2 gauss.)

As we shall see next, a neutron star is known to exist inside the Crab Nebula. But is presence is detected not through its oscillations, as we had suggested, but through its rotation: for a pulsar is not an oscillating but a rapidly rotating neutron star.

THE GOLD MODEL OF A PULSAR

Let us now return to the finding of Antony Hewish and Jocelyn Bell. They had found rapid pulsation and the question was, what type of object could be small enough to serve as its source. In 1968, theoreticians had two possible candidates, the white dwarf and the neutron star, and a number of different theories sprang up to explain the nature of pulsars. In the early days after the discovery of CP 1919, a few more pulsars were found, thus providing further checks and constraints on the theories; a number of these fell by the wayside in the usual scientific competition for the survival of the fittest. In particular, it became clear that the white dwarf could be ruled out and the neutron star, of much smaller size, was the more likely source. Likewise, the cause of the pulses was found to be not the oscillations of the star but its fast spin.

A model proposed by the Cornell astrophysicist Tommy Gold (see Figure 4.6) in 1968 ultimately emerged as the best buy from among these theories. And although today we still do not have a very detailed model of pulsars, the Gold model serves as a good starting point for any more elaborate exercise to understand them. What may be going on in and around a neutron star can be understood according to Gold's scenario in the following way.

Figure 4.6:
Tommy Gold.

A neutron star has two polar axes: a rotational axis and a magnetic axis. The Earth also has two sets of poles, one from its rotation axis and the other from the magnetic axis. But unlike the case of the Earth, where the two axes are nearly aligned, in a typical neutron star the two axes may be pointing in very different directions.

The rotating star has a swarm of electrically charged particles (the electrons) in its atmosphere. As the star rotates so does the atmosphere, carried along by the star's strong gravitational pull. Just as the outer parts of a merry-go-round move much faster than its inner parts, the charged particles in the outer parts of the atmosphere move very fast, and may approach the speed of light. For a pulsar spinning once per second, this limit may be reached at a distance of around 50 000 km from the axis of spin. Such fast particles are known to radiate electromagnetic waves in the presence of magnetic fields. The radiation is highly beamed like the beam of a rotating searchlight. See Figure 4.7 for a schematic picture of this model.

So if we happen to be located in the sweep-through area of the pulsar beam we will get pulses of radiation each time the beam sweeps past us. The pulse period is therefore just the period of rotation of the neutron star about its axis.

If we follow the Gold model further, we may ask the question: what happens to the spinning neutron star as it keeps on radiating for a long time? Obviously, the process cannot go on for ever. Indeed, as time goes on, the spinning pulsar slows down and its pulse period increases. Thus we can imagine that the pulsar starts off spinning very fast and, as it ages, it slows down. A pulsar which has a pulse period of one second today may slow down to a two-second period after, say, a million years.

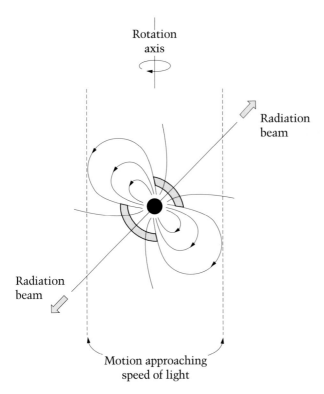

Figure 4.7: The Gold model of a pulsar. The magnetic field lines start and finish at the central neutron star, which is surrounded by a band of charged particles. As the star spins the charges move across the magnetic field lines resulting in radiation along the magnetic axis.

Looking at pulsars of different periods, therefore, the astronomer can tell broadly which pulsar is old and which one has just been started off. The magnetic field also decreases as the pulsar ages and this is a factor in the change in the intensity and spectrum of its radiation.

However, although this picture seemed to rest on reasonably secure foundations, there were further surprises in store for pulsar observers, as we shall see in the later parts of this chapter.

THE CRAB PULSAR

Looking at the sequence of events leading to the formation of a pulsar, we note that first a star has to explode, casting off its envelope. It leaves behind a spinning core which becomes a rapidly spinning neutron star. Assuming that it also has a magnetic field, we expect it to become a pulsar.

Following this line of argument we should see a pulsar near the remnant of a supernova. An ideal case therefore would be that of the Crab Nebula. Indeed, a pulsar was found in the Crab Nebula. It was the second pulsar to be found. Its discovery also provided the resolution of a long-standing puzzle about the Crab.

Figure 4.8 shows another photograph of the Crab Nebula, which we saw earlier in Figure 3.1. We see the remnant left over after a star exploded about nine and a half centuries earlier. The nebula is a scene of various kinds of activity indicating that very energetic processes were still going on there at the epoch we are observing. For example, apart from the optical wavelengths, the Crab is known to radiate strongly through radio waves, as well as in X-rays and gamma rays. Let us first pause and take cognizance of these different forms of

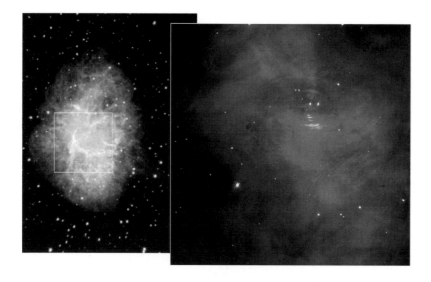

Figure 4.8: The Crab Nebula. On the left, a ground-based telescopic image. On the right, the central part imaged by the Hubble Space Telescope, with the Crab Pulsar visible as the left of the pair of stars near the upper centre of the frame (courtesy of Jeff Hester and Paul Scowen, Arizona State University, NASA and STScI).

radiation. The following two paragraphs summarize what we have already discussed in Chapter 1.

Scientists know that light is an example of wave motion, the waves being formed from electric and magnetic disturbances of an undulating nature (see Figure 4.9). Just as we see an undulating water surface with waves travelling outwards when we throw a stone into a pond, so do electromagnetic waves travel outwards from a source of light. Figure 4.9 explains what is meant by the wavelength of a wave.

What we are familiar with as the visible form of light (that is the light to which our eyes react to enable us to 'see' things) has a wavelength range approximately between 390 to 770 nanometres, a nanometre being a thousand millionth part of a metre. What does the wave represent if its wavelength lies outside this range? Broadly speaking, we divide the entire wavelength range into several regions, the one containing the longest waves being labelled the radio-wave region and that containing the shortest being the gamma ray region. In between lie microwaves, infrared radiation, visible light, ultraviolet radiation and X-rays, in decreasing order of wavelength (see Figure 4.10).

Astronomical objects emit radiation in the form of electromagnetic waves, of which the most familiar to us is, of course, visible light. But, as we have seen, they also emit radiation at other wavelengths; in some cases, far more than they radiate in visible light.

The Crab Nebula is one of such cases. The radiation in radio or X-rays requires the supply of fast-moving electrons in the ambient magnetic field. We expect such electrons to be around in the nebula; however, there is a difficulty.

Recall from Chapter 3 that the explosion in the supernova would have released a large number of fast-moving particles including electrons. We considered this as a possible source of cosmic rays. But the

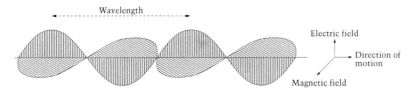

Figure 4.9: An electromagnetic wave is visualized here with wave-like forms showing how the electric and magnetic disturbances in perpendicular planes rise and fall in consonance through space. The distance between two successive peaks of the wave is called its *wavelength*.

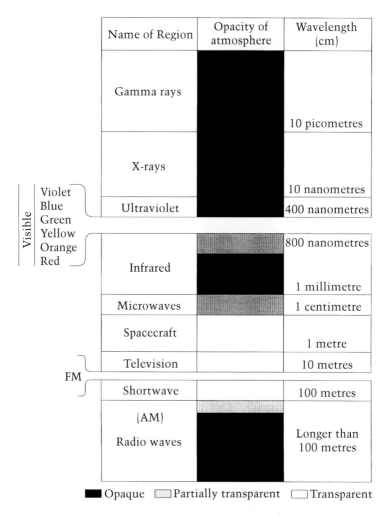

Name of Region	Opacity of atmosphere	Wavelength (cm)
Gamma rays		
		10 picometres
X-rays		
		10 nanometres
Ultraviolet		400 nanometres
Infrared		800 nanometres
		1 millimetre
Microwaves		1 centimetre
Spacecraft		
		1 metre
Television		10 metres
Shortwave		100 metres
(AM) Radio waves		Longer than 100 metres

Visible: Violet, Blue, Green, Yellow, Orange, Red

FM

■ Opaque ▨ Partially transparent ☐ Transparent

Figure 4.10: This chart, reproduced from Figure 1.13:, shows the different wavelength ranges of electromagnetic waves.

explosion was a one-shot affair. Even if the electrons released at this time, nine and a half centuries before the time at which we observe it, were still present in the nebula, they would have lost most of their energy of motion and would have slowed down. It was therefore a mystery how the *presently seen* radiation was being produced. This was the problem that bothered many astrophysicists.

In a personal anecdote about the Crab Nebula, Fred Hoyle recalls that in 1958 he raised this problem at a private discussion during the Solvay Conference at Brussels, when the senior Dutch astrophysicist Jan Oort and the astronomer Walter Baade were present. Baade had been instrumental in detailed studies of the Crab and Hoyle had asked him if it would be possible to look for a source in the Nebula. Baade wanted to know what exactly he should look for. Basing his idea on a pulsating white dwarf, Hoyle suggested a source whose light fluctuates over a period of a few seconds. Although interested in looking for such a source, Baade did not follow it up, probably because the photographic techniques available to him were not sensitive enough.

In the end the source was discovered in 1968 by D.H. Staelin and E.C. Reifenstein at the National Radio Astronomy Observatory at Greenbank in the United States. Actually, the first discovery was only of some isolated 'giant' pulses which are occasionally emitted by the source. Later investigations revealed the source to be a pulsar with a very short period, only 0.033 second, or 33 *milliseconds*. (It is often convenient to use the shorter time unit of a millisecond, that is, a thousandth part of a second, to describe the pulse period of a fast pulsar.)

However, apart from this very fast-pulsating radio source, the Crab had other surprises in store. On 16 January 1969, the discovery was made of *optical* pulses from the Crab pulsar. The actual discovery was found on a tape recorder that had been accidentally left running by the observers William Cocke, Mike Disney and Donald Taylor at the Steward Observatory in Tucson, Arizona. Subsequently, two more groups reported the finding of optical pulses, one from McDonald Observatory, Texas and the other from the Kitt Peak National Observatory, also at Tucson. The sequence of photographic frames in Figure 4.11 shows how the pulsar image alternately brightens up and fades out. The curve shown below the photographs illustrates how the rise and fall of visible intensity makes up a typical couple of pulses.

The next interesting addition to the story came in the same year from the budding science of X-ray astronomy. Two rocket flights with X-ray detectors, one from the Massachusetts Institute of Technology and the other from the US Naval Research Laboratory, showed that even in X-rays the source pulsates. And the shape of the pulses in X-rays matched reasonably the shape of their optical counterparts.

The emission in the optical and X-rays from the Crab pulsar is pulsed like the radio emission, and is therefore in the lighthouse-beam mode as described above. However, it originates high above the pulsar surface, in

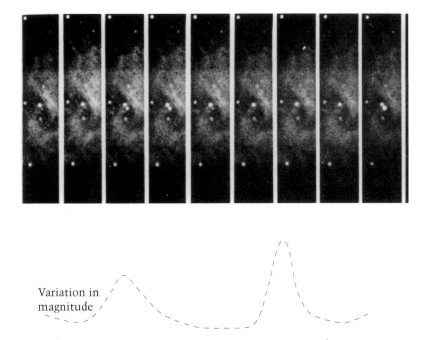

Figure 4.11: The sequence of photographs illustrates the visual brightening and fading of the source in the Crab Nebula. Below we see the rise and fall of the intensity plotted on a graph (courtesy of S.P. Maran, Kitt Peak National Observatory).

the magnetized atmosphere where the electrically charged particles are moving very close to the speed of light. For, to get radiation at the high frequencies of the optical or X-ray regions (as opposed to the much lower frequencies of radio waves) the charged particles need to have energies as much as a billion times their rest-mass energy.[5]

The pulsar in the Crab Nebula is called NP 0532 (or, PSR 0531+21 in the more standard format) and it is believed to be the primary source of energy in the nebula. However, it is not the general case that you will find a pulsar in or near any supernova remnant. The reason is that a

---

[5]This energy E is given by Einstein's equation $E = mc^2$, where m is the rest mass of the particle and c is the speed of light.

supernova may explode in a skew fashion (see Figure 3.21 of the last chapter), throwing off the remnant core away from the envelope. So the Crab pulsar is something of an exception in being found within the site of the explosion.

We now leave the Crab Nebula and its remarkable energy machine to look at some novel aspects of the pulsar phenomena that came to light much later than the original discovery.

BINARY AND MILLISECOND PULSARS

From what has been said so far it may appear that pulsars are necessarily born from the core left behind after a supernova explosion. Such pulsars would start spinning rapidly but will gradually slow down and also suffer a decay of their magnetic field. Indeed, the rate at which the pulse period of a pulsar increases can be linked to its emission mechanism, and we arrive at the picture that the older the pulsar the slower is its spin. A crude but simple rule of thumb that gives us the 'age' of the pulsar is: divide the observed period (repeat time) of the pulses by double the rate at which the period is observed to decrease. The answer is a good approximation to the age of the pulsar.

Figure 4.12 shows a plot of pulsars whose period (plotted on the horizontal axis) and rate of increase of period (on the vertical axis) are both known. Such a diagram is useful in understanding how a pulsar evolves with increasing age, just as the H–R diagram is useful in understanding the evolution of stars. Notice that a large number of pulsars are clustered in the upper right part of Figure 4.12. These fit in with the supernova scenario described above.

However, there are a few pulsars which have very low periods and a still lower rate of increase of period. Some of these (circled in Figure 4.12) are in binary systems. Their ages, by our formula above, are in the neighbourhood of *a billion years*. Looking at Figure 4.12 one might feel that these are a different breed altogether! Indeed they are, and the clue to their origin lies in the way a binary star system evolves.

A binary system is made up of two stars going round each other. We see them quite commonly in the sky, although they are difficult to spot with the naked eye. In some cases, however, the two stars are quite close to one another and this leads to an exchange of mass between them. Thus it may happen that one star is a very compact neutron star while the other is a huge giant. The former may then be able to

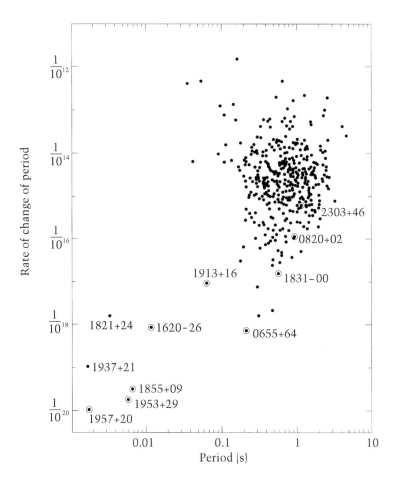

Figure 4.12: This diagram has a large number of 'standard' pulsars in the top right part. These fit in with the theory that their origins are in supernovae. But what do we make of those in the bottom left corner? The circled pulsars are in binaries.

pull matter from the surface of the latter; this matter then travels fast and falls into the compact member. As it falls, however, it circles around it, spiralling inwards. Figure 4.13 shows such an arrangement. The spiralling matter heats up by friction and radiates X-rays. The advent of X-ray satellites in the 1970s revealed the existence of several such binary sources of X-rays. We will return to this picture in the following chapter.

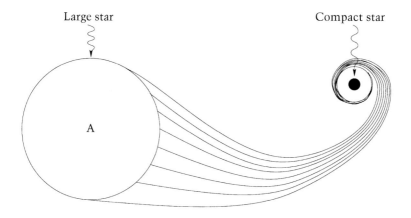

Figure 4.13: The scenario of a binary X-ray source as described in the text.

This scenario is found to be a common one for binary X-ray sources and, before considering how it leads to pulsars, it is interesting to look at how the binary star system itself evolves to this stage. Figure 4.14 illustrates the typical evolutionary sequence in four stages. We start at stage (*a*) with a pair of stars *A* and *B* with masses 8 and 20 solar masses respectively. Star *B*, being more massive, evolves more rapidly. After 6.2 million years, *B* becomes a giant star and acquires a radius so large that it cannot hold itself together under the tidal forces exerted by its companion.

The word 'tidal' arises from the example of ocean tides. The extra gravitational pull exerted by the Moon on the surface of the Earth facing it causes the oceans in that region to rise, thus leading to high tides. The same happens to the outer parts of *B* because of the pull from companion *A*. As a result the matter from *B* begins to flow towards *A* in the manner shown in stage (*b*). The dotted horizontal figure of eight shown in (*a*) and (*b*) is the so-called *Roche lobe* (named after E. Roche who first pointed out in 1850 the possibility of tidal disruption exerted by a planet on its satellite). This lobe decides the maximum extents of the two stars for them to be able to retain undisrupted shapes. Once a star swells beyond its Roche lobe, it starts losing its surface matter. After 6.8 million years, star *B* explodes as a supernova leaving behind a neutron star. The mass of star *A*, meanwhile, has grown to 24 solar masses as a result of accretion of matter from star *B*. This is shown in stage (*c*). Finally stage (*d*) brings us to the characteristic situation of a binary X-ray

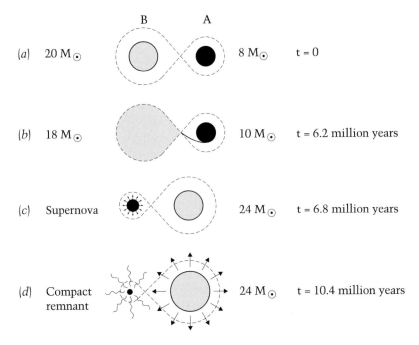

(a)	20 M$_\odot$	8 M$_\odot$	t = 0
(b)	18 M$_\odot$	10 M$_\odot$	t = 6.2 million years
(c)	Supernova	24 M$_\odot$	t = 6.8 million years
(d)	Compact remnant	24 M$_\odot$	t = 10.4 million years

Figure 4.14: Four stages in the evolution of a binary star system in which exchange of mass between the two components takes place. $M_\odot$ denotes the mass of the Sun.

source. In this situation star A has become a supergiant and matter from its surface crosses the Roche lobe and begins to flow to star B. This 'stellar wind' is the cause of X-radiation in the manner discussed before.

In this way we see that in a binary system mass transfer plays a key role. Since the mass which moved over from one star to another was orbiting the common centre of mass, when it fell onto the second star it carried to it this tendency to rotate. As a result the second star will spin faster. Thus we expect that the pulsar left over after one member of the binary (B, in our example) becomes a supernova, will spin very fast.

This explains the so-called 'millisecond pulsars', whose periods are measured in milliseconds rather than in seconds. One such fast milli-second pulsar, PSR 1957+20, has a period of only 1.6 milliseconds. It was discovered in 1988 by Andrew Fruchter, D.R. Steinberg and Joe Taylor.

Returning to binary evolution, the ultimate fate of the two members could be that both become supernovae, leaving behind two neutron stars. Alternatively, the system could be blown apart leaving behind one neutron star. Thus it is possible to have fast-spinning pulsars as singletons or as members of binary systems.

The binary pulsar PSR 1913+16

The most famous of the binary pulsars, and the first one to be found in that category, was discovered in 1974 by Russell Hulse and Joe Taylor at the 1000-foot radio dish in Arecibo in Puerto Rico (Figure 4.15). Known by the catalogue designation PSR 1913+16, this pulsar, together with another neutron star, moves in a binary mode with the very short orbital period of approximately $7\frac{3}{4}$ hours. Both stars are of mass approximately 1.4 solar masses. The pulsar has a short period of 59 milliseconds and also a slow rate of increase of the period. The period is very stable and can serve as a clock with an accuracy of 50 microseconds if we average the pulse arrival times over a period of five minutes.

Figure 4.15: The dish has been dug into the ground and receives radio signals from a limited belt on the sky as Earth spins about its axis. Its design is particularly suited for finding pulsars (courtesy of NAIC/Arecibo Observatory).

We will now come to this timing accuracy and also to the remarkable use to which PSR 1913+16 has been put by physicists for testing theories of gravity. For their discovery of this remarkable pulsar, Hulse and Taylor were awarded the 1994 Nobel Prize for physics.

PULSARS AS STANDARD CLOCKS

We have already remarked on the fact that the time period of the first pulsar, CP 1919, could be quoted to ten decimal places. The unusually steady periods of pulsars, especially the millisecond pulsars which were discovered in the 1980s, opened up the possibility that pulsars could serve as the basic time keepers for natural phenomena.

The present definition of Universal Time (UT) is in terms of the idealized caesium clock. This clock depends on the oscillations of the caesium atom. In practice the second is defined as the duration of 9 192 631 770 periods of the radiation corresponding to the transition between two specified states of the caesium atom. The characteristic time intervals associated with each such atomic transition are, however, not strictly the same. But, by averaging over several such clocks, one can arrive at a steady time period. Pulsars, however, seem to do better in terms of providing us with a steady time standard, as can be seen below.

The quantitative estimate of the steadiness of a clock is provided by the so-called Allan variance of its errors. To obtain this variance, one measures the fluctuations in time period as a fraction of the latter and averages the squares of such fluctuations. This variance decreases if we can measure it over a longer interval of time, provided we are confident that the basic time period remains stable over the interval of measurement. Thus the longer this time interval, the less is the Allan variance and the more accurate is the clock.

For a caesium clock the interval is of the order of a month. In Figure 4.16 we see how the variance falls over a period of a million seconds or so and then begins to rise. By contrast, the same figure shows that for the pulsar PSR 1937+21 the time interval of steady period *runs into years*! That is, over short time scales of a month or so the pulsar may not do as well as the atomic clock. But for longer durations its stability tends to overtake the latter, making it accurate to *thirteen decimal places.*

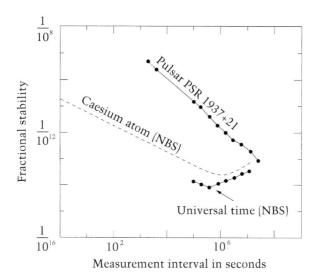

Figure 4.16: This graph shows the comparative behaviour of the caesium clock and the pulsar PSR 1937+21. The horizontal axis plots the time interval over which measurements are taken while the vertical axis plots the fractional stability of the clock, as measured by the Allan variance.

Thus one could proceed as follows to construct a purely pulsar-based time standard. Suppose one could show, by making observations of the smoothest running pulsars, that the difference between the time standards given by two pulsars is *less than* the difference between UT and the mean pulsar standard. If so, one could rely entirely on pulsars as the basic timekeepers. This would certainly improve the UT standard by bringing down the fluctuations within which it is currently measured. Whether pulsars will ultimately replace atomic clocks is an open question.

PULSARS AND THE TESTS OF GRAVITY THEORIES

The very accurate time keeping by pulsars has helped physicists in another way.

The binary pulsar PSR 1913+16 introduced earlier has proved to be extremely useful in testing the predictions of Einstein's general theory of relativity vis-à-vis some other gravity theories. We will not go into the details of relativity here. The reader may find a description of the theory[6] in Chapter 5.

The general theory of relativity has a starting point very different from the much simpler Newtonian law of gravitation but for most practical purposes it ends up giving the same answers. Thus, to test which of the two theories is closer to reality one needs more subtle tests that require very precise measurements and rather special circumstances. Such tests, within our own solar system, have been primarily responsible for increasing the credibility of general relativity at the expense of Newtonian gravitation. These tests, however, require very accurate measurements.

The advance of periastron

For example, take the test provided by the motion of the planet Mercury around the Sun. Figure 4.17 shows that according to Newtonian gravitation Mercury should move in an elliptical orbit with the Sun as a focus of the ellipse.

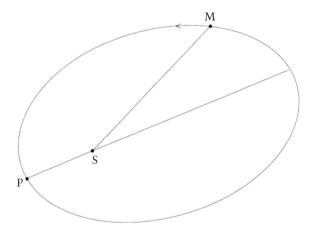

Figure 4.17: The basic motion of Mercury (M) under Sun's (S) gravitational field is an ellipse with the Sun at the focus. Notice that the distance of the planet from the Sun changes continuously, being shortest when the planet is at P, the *perihelion* point.

[6]See also the author's book *The Lighter Side of Gravity*, Cambridge University Press, for a longer non-technical account.

However, in practice, the motion of Mercury is a little more complicated, as shown in Figure 4.18. The line joining the Sun to the *perihelion*, the closest point in the orbit, slowly changes direction with time.

This strange behaviour had been noticed in the last century and several attempts were made to understand it within the framework of Newtonian gravitation. Thus a substantial fraction of this movement of the line *SP* in Figure 4.18 was known to be due to the gravitational pull on Mercury by other planets of the solar system, especially Venus, Earth and Jupiter. Nevertheless, a tiny balance remained unaccounted for.

How tiny was the anomaly can be seen as follows. Figure 4.19 shows the type of protractor used in a school mathematics kit. It measures angles in tiny divisions marked on its circular boundary. Each division is a degree. Divide the degree into 60 equal parts to get an even tinier measure of angle called a *minute of arc*. Next make 60 divisions of a minute of arc to get a *second of arc* (an arcsecond). The anomalous shift of Mercury's perihelion as viewed from the Sun was at the rate of *43 seconds of arc in 100 years*.

Tiny though this discrepancy looks, it was sufficient to worry the theorists, who up till then had found Newton's law of gravitation in complete conformity with observation. And it was here that general relativity stepped in with the right answer. It introduced a small mod-

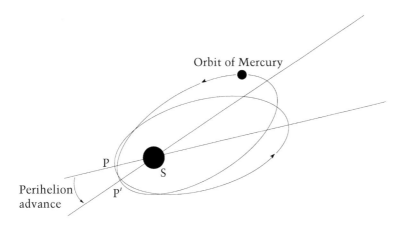

Figure 4.18: The line *SP* joining the Sun to the closest point of Mercury's orbit slowly rotates in space. The effect is shown in an exaggerated fashion above.

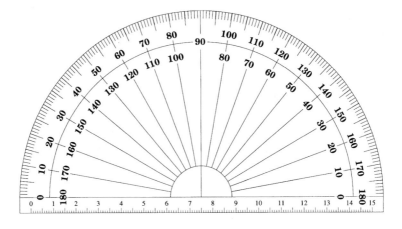

Figure 4.19: This picture of a protractor reminds us how tiny a one-degree angle is.

ification in the way a planet moves round the Sun and showed that it *exactly* accounted for the anomalous 43 seconds of arc per century.

I have, apparently, disgressed from pulsars to planets, to show how tiny, yet significant, was the difference between two gravitational theories, one of Newton, the other of Einstein. It is against this backdrop that we have to view the enormous improvement in time measurement provided by binary pulsars.

In Figure 4.20 we see how the two stars in PSR 1913+16 move in a binary system, each following an elliptical orbit. The line joining them, however, passes through a fixed point in space called the *centre of mass*

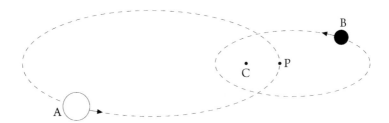

Figure 4.20: The pulsar *A* and the companion star *B* move in elliptical orbits such that their centre of mass *C* is fixed in space. The pulsar is at periastron *P* when the distance *AB* is the least. It is observed that the direction *CP* changes with time.

of the pair.[7] Of course, while the centre of mass stays constant, the distance between them varies. Like the perihelion in the case of the Sun, we may talk of a *periastron* for binary stars when the distance between them is least.

Actually the Sun–Mercury system could also be viewed as a binary system. The mass of the Sun, however, is some six billion times the mass of Mercury. The result of this huge ratio of masses is that the Sun hardly budges under Mercury's tiny force of attraction: their centre of mass is *almost* coincident with the centre of the Sun.

This particular circumstance enables the relativist to calculate the rate of advance of the perihelion of Mercury almost *exactly*. In the case of a binary pulsar, the situation is different. The two stars (pulsar *A* and companion *B*) are of comparable mass and so a repeat of the calculation in exact terms is not possible. The so-called *two-body problem*, in which two comparable masses move under each other's gravitational pull, is extremely difficult and *has not been solved in general relativity*.

Nevertheless, one can make approximate calculations which are considered credible by pundits in the field. These calculations give a value for the advance of periastron of PSR 1913+16 of the right magnitude, 4.2 degrees per year, as observed. (Notice that this effect is about 350 000 times that observed for Mercury.) The binary pulsar therefore provides a confirmation of general relativity through the observed shift of periastron.

Time delay

Another effect peculiar to general relativity (and not found in Newtonian gravitation) is to do with the time delay in a light signal passing close to a massive body. We will see in the next chapter how general relativity requires space–time measurements to be modified in the neighbourhood of such a body, because of its gravitational influence. Thus the to-and-fro passage of a radar signal would have a longer duration if such modifications were present.

In the solar system this effect was observed by the Mariner spacecraft in bouncing radio signals off the surface of Mars, *when these signals grazed the Sun*. Compared to the situation when the Sun was nowhere

[7]Imagine two children sitting at the ends of a see-saw which is horizontal. In such a situation, their centre of mass is the point where the see-saw rests. If one child is very heavy, he or she will have to move in closer to this point in order to keep the see-saw in balance.

near such signals, the delay was about 250 microseconds (see Figure 4.21).

In the case of the binary pulsar PSR 1913+16, the pulsar signal will take about 50 microseconds longer to reach us when it grazes the pulsar's companion. The effect, though small, can be accurately measured, thanks to the precise time keeping of the pulsar. And the measurements have confirmed the above prediction of general relativity.

Existence of gravitational radiation

Although these tests have been very fruitful in pushing up the credibility of general relativity one notch higher, none of them has generated as much excitement as the reported (indirect) evidence for the existence of gravitational waves. Let us first look at how these waves are expected to be produced.

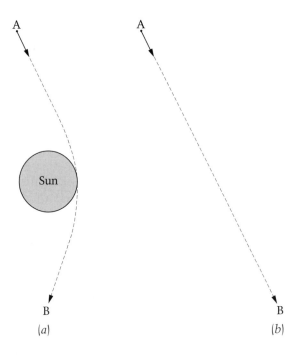

Figure 4.21: (a) Schematic picture of a signal that grazes the Sun's surface. Its time of passage from A to B is longer than when the Sun is not in the picture, as in (b).

An analogy with electromagnetic waves will help. The most basic mechanism for the emission of such waves is an oscillating electric charge. The to-and-fro motion of such a charge will generate energy in the form of electromagnetic waves (see Figure 4.22). A detector of electromagnetic waves can easily measure this radiation. However, we can indirectly infer its existence by asking the question: Where is this energy coming from? It has to come from the motion of the electric charge. Therefore, as the charge continues to radiate, its motion slows down, much like a car slowing down against friction of the road when its engine is turned off. So, from the damping of its motion, we can deduce that the electric charge has been radiating energy.

Just as an oscillating charge radiates electromagnetic waves and slows down as a result, so do we have massive dynamical systems radiating gravitational waves and slowing down. In theory, that is, for no one has yet succeeded in directly detecting gravitational waves.

General relativity tells us that the simplest system radiating gravitational waves would be a binary system, where two masses go round

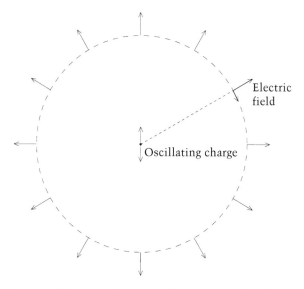

Figure 4.22: The arrows show an oscillatory motion of an electric charge. The charge will radiate electromagnetic energy at large distances, shown by the outward arrows across the dotted circle. Such emission occurs in the planes passing through the line of motion of the charge. The electric field is typically in such a plane while the magnetic field is perpendicular to it, both being perpendicular to the outward direction of the wave.

each other, as in the case of the binary pulsar PSR 1913+16. It is expected that because of the energy lost in gravitational radiation, the binary pulsar system should shrink. That is, the two members should move around each other in smaller and smaller orbits. As the orbits shrink, the binary period *decreases*. The theoretically estimated decrease is as small as 2.4 picoseconds per second! (One million million picoseconds make one second.)

Nevertheless, thanks to the accurate timing kept by the pulsar, this minute effect has been measured and verified. One can see this by observing the cumulative change of phase in the orbit, which was two seconds over six years. Figure 4.23 shows a graph of such observations.

The fact that the orbital period has decreased is apparent from this phase change and the steady rate of decease is seen as a confirmation of the general relativistic prediction that such binaries should radiate gravitational waves. There are other theories making similar predictions, although of different amounts, but the accuracy of measurement of this binary system is such that it rules out such alternatives.

We may have to wait until early in the twenty-first century for a direct proof that gravitational waves exist. Several large detectors are

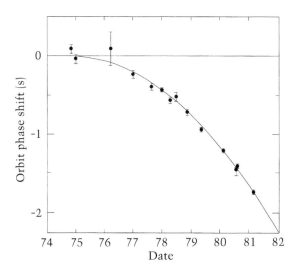

Figure 4.23: The relative positioning of the two members of a binary system can be quantified by the *phase* of the orbit, measured in seconds. As the orbit shrinks and the two members move faster round each other this phase changes. The phase change was measured for PSR 1913+16 and is given here as plotted by J.H. Taylor and J.M. Weisberg in 1984.

currently under construction with the specific aim of picking up gravitational waves by binary stars in orbits that progressively shrink until the stars coalesce. The binary pulsar, however, reassures us that such waves do exist!

PLANETS AROUND PULSARS

We discussed earlier how planets form whenever a star is born (see Chapter 3), through the shrinkage of a gas cloud in interstellar space. Normally, therefore, one expects to find planets around stars like the Sun, which have been steadily producing energy through the fusion of hydrogen to helium. Indeed a few cases of such stars with planets are now known.

In 1991, however, there was a claim that a planet had been found around a pulsar! Keeping in view the somewhat traumatic origin of a pulsar, how can it manage to acquire a planet? Surely, any planets that a star may have had before it became a supernova would have been blown away or destroyed by the explosion. Therefore, when, in 1991, a group of radioastronomers from Jodrell Bank announced that a particular pulsar did seem to have a planet around it, the news came as a complete surprise.

How did the astronomers make this discovery? The pulsar signals seemed to show a small wobble that could only be explained if the pulsar had a planet around it, gravitationally disturbing it. The situation is somewhat similar to that for binary stars, where each star affects the motion of its companion; but, in this case, the planet being far less massive than the star manages to produce only a barely perceptible effect. Thus the star will wobble slightly as the planet orbits it. The extent and period of the wobble, if measurable, can tell us something about how massive the planet might be and how long it takes to orbit the star. (Remember, the planet itself, being non-luminous, is not visible.) Thus the Jodrell Bank group was relying on this indirect evidence for making their claim.

The announcement of the discovery was, of course, an immediate sensation. As is not uncommon when such unexpected findings are announced, a special conference was later arranged to discuss the details and implications of this discovery. But the discovery itself turned out to be a false alarm! Suspicion as to its validity in fact arose when it was discovered that the predicted planet seemed to have a period of six months or a year, exactly matching the Earth's period! Ultimately it turned out that because we are observing the pulsar from the moving Earth, our motion also affects the data and produces the periodic pattern. So, this was no real effect: it was simply the result of observing a pulsar from a moving platform. Ironically, at the conference at which this discovery was retracted by Andrew Lyne from Jodrell Bank, Aleksander Wolszczan, an astronomer working at the radio telescope at Arecibo in Puerto Rico, reported that *he had found a pulsar with two planets.* The pulsar had the catalogue number PSR 1257+12.

Now, if you have had a false alarm once you are less inclined to believe a similar case. You will obviously ask for a check and double-check of the records. But Wolszczan had already taken sufficient care to see that the effect of the Earth's motion was accounted for and that any other spurious effect was removed. Thus he was confident of the reality of the effect, which was also double-checked by others.

Thus, at least two planets are known to move around this particular pulsar; one has a mass of 2.8 times the mass of the Earth and the other a mass of 3.4 times the mass of the Earth. Their respective periods around the pulsar are 66.6 days and 98.2 days. So they are moving relatively fast, like Venus or Mercury. Their respective distances from their parent star are 70 million kilometres and 54 million kilometres. That is, they are relatively close by. (Just for comparison the Earth is orbiting the Sun at a distance of 150 million kilometres.) Now observers have claimed that there is a third planet also in that system. But we still do not know how those planets got there. This is a problem for the theoreticians to worry about!

UNFINISHED STORY

There we conclude our narrative of the fourth wonder of the cosmos. There are still many questions to answer about pulsars. Suffice it to say that the field of pulsars has continued to add new dimensions to the original discovery of 1967. As Taylor and Stienberg concluded in a review article: *The field is ripe for new ideas and new enthusiasm.*

Gravity: the great dictator ⑤

A MOMENT OF BRAHMA

The Indian scripture *Bhagavatam* has the following story about Brahma, the first of the trinity of Gods, who creates the universe. A king called Kukudmi had a beautiful daughter called Revati, who attracted a large number of suitors. Anxious to make the right choice of son-in-law, Kukudmi decided to consult no less an authority than Brahma. So he went to Brahma's abode along with his daughter. As he was occupied with some important matter of the universe, Brahma sent word to Kukudmi to wait for a moment.

Presently Brahma came out, and having greeted the king asked the reason for his visit. Kukudmi explained his problem and sought advice from the Great One. Brahma laughed and said: 'Let me explain. All those suitors on your list are no more. They are dead and gone. Although you waited for a moment in my abode, during that period, thousands of years have passed on the Earth.' The king was naturally dismayed. However, Brahma offered a solution: 'Never mind. Go back to your kingdom', he said. 'There you will find a suitable boy called Balarama for your daughter.' And Brahma gave Kukudmi details of this suitor-to-be.

This is one of the oldest stories that I know of wherein the concept of time passing at different rates for different people or places plays a key role. In another context, Hindu mythology defines one night of Brahma as equal to 4 320 000 000 years on the Earth. We will find echoes of this time scale in modern cosmology in Chapter 7.

Coming back to our daily experience of modern times, the concept of time flowing differently for different observers seems bizarre and the above tale appears just an impractical notion made up into a folklore.

But, considering the march of ideas about time and space in the twentieth century, the scientists may not find the story all that strange.

Thanks to the revolution brought about by Albert Einstein (Figure 5.1) through his theories of special and general relativity, we should not be surprised at all by the Brahma story. Indeed, astronomers encounter examples in the cosmos which would match it. These are examples where one of the basic forces of nature plays a key role.

That force is the force of gravity.

SPACE, TIME AND MOTION

To relate the tale about Brahma to modern concepts of space and time we go back to 1905, when a young worker in a Swiss patents office in Berne wrote a paper that revolutionized those concepts. Albert Einstein's paper was called 'Electrodynamik, bewegter Körper' (electrodynamics of moving particles). Why did it introduce radically new ideas on how observers measure space and time?

Let us begin with a familiar-looking example of measurements in space. In Figure 5.2 we have a city with streets and avenues well laid out in a rectangular fashion; we will say that streets go north–south and avenues go east–west. Suppose that there are two locations, A and B, in the city and that we wish to measure the distance between them *in a straight line, as the crow flies.*

Naturally this is not so simple, since humans cannot fly as crows. Nor can they cut across walls and yards through which the straight line goes. They are constrained to walk along the streets and avenues. So we go east along the avenue through A as far as location C and then north along the street through C as far as B. We can measure the distances AC

Figure 5.1:
Albert Einstein.

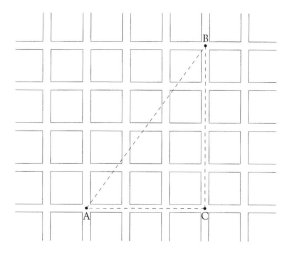

Figure 5.2: A city with a network of streets and avenues laid out in a rectangular fashion in the north–south and east–west directions. How do we measure the distance AB? We can, of course, measure AC and CB.

and CB, and then draw the triangle ABC as shown in Figure 5.3. Knowing that the angle ACB is a right angle, we can compute the distance AB by using the theorem of Pythagoras $AB^2 = AC^2 + CB^2$.

For example, if AC is 3 kilometres, and CB is 4 kilometres, then the above theorem tells us that AB will be 5 kilometres. So we can in general work out the distance AB by measuring the two separate sections AC and CB in two perpendicular directions.

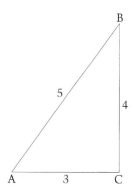

Figure 5.3: Triangle ABC has a right angle at the vertex C.

Let us now consider a slightly different situation, shown in Figure 5.4, where our city does not have avenues and streets in the east–west and north–south directions; rather, in this case the avenues are from southwest to northeast while the streets are from northwest to southeast. Thus the new path system is obtained by turning the earlier system around by 45 degrees.

Naturally, if we repeat our experiment of going via a point C which is located on the avenue through A and street through B, we will have different lengths for AC and CB and our new triangle will look like that in Figure 5.5. *However, so far as the distance AB is concerned it is the same.* Our application of Pythagoras' rule to the new triangle will give us the same answer, even though the street–avenue system is different from before, having been rotated by an angle of 45 degrees.

In mathematical language we say that *the distance AB is invariant under the rotation of the path system.*

Thus, the distance AB has a somewhat special status, vis-à-vis the other two lengths AC and CB. No matter how much we turn the city's path system around, the distance AB will stay the same, although the other two distances AC and CB will be different each time.

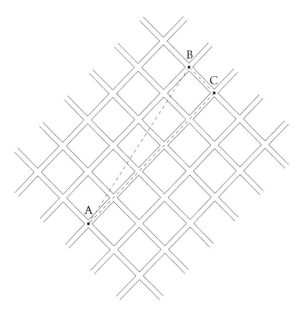

Figure 5.4: The city of Figure 5.3 but with its streets and avenues rotated by an angle of 45 degrees.

GRAVITY: THE GREAT DICTATOR

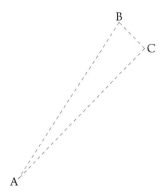

Figure 5.5: The triangle ABC has a right angle at C and though its sides BC, CA do not equal the corresponding sides of the triangle ABC of Figure 5.3, the sides AB of the two triangles are still the same.

The example of the rotating path system illustrates the basic feature of a two-dimensional space. To locate any intersection in the city we need two directions, viz. the street and the avenue on which it is located. The number of dimensions of space thus equals the number of items of information needed to specify any location in that space. For example, imagine that at the location B there is a skyscraper and we need to find someone staying there. For that purpose we need to specify the floor on which the person stays. Thus we need *three* items of information, because we are now dealing with a three-dimensional space. See Figure 5.6, where the person's abode is shown at D.

However, the basic property of invariance of the distance between two points in space continues to hold in three dimensions. No matter how we specify the three items of information needed to get from A to the person at D, the distance AD will always be the same.

This is easy enough to understand, and was known well before Einstein came on the scene. In fact from Isaac Newton's days, scientists have been accustomed to describing a location in the real world with three *coordinates*, that is, three items of information. In addition, if they wished to label an event taking place at the location, they needed *one more* item of information, namely, *when* did the event take place. This additional item of information is the time coordinate.

Imagine an accident in which a person crossing a road is knocked down by a passing car. For the accident to take place, it is essential for the person to be at the same place as the car, *at the same time*. Unless all the four coordinates match, the event of collision would not take place. Hence the real world of events is of *four dimensions*, three of space and one of time.

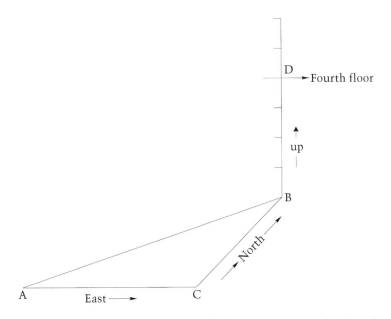

Figure 5.6: To locate a person living at *D* in the skyscraper at *B*, we need additional information, viz., the floor on which *D* is situated.

However, as we all intuitively feel, the fourth coordinate, time, is somewhat different from three space coordinates. After all, we measure all the distances in Figure 5.6, along the street, along the avenue and up the skyscraper, in metres but we measure time in hours, minutes and seconds. For spatial measurements we use the metre rod, while for the temporal one we use a clock. Thus space and time, although both needed to specify completely the 'where' and 'when' of events, stand apart.

This is what Newton believed when he assigned an absolute status to space and an absolute status to time. The clocks of observers located anywhere and moving in different directions with different speeds would record time in the same way. Similarly all these observers would measure spatial separations in the same way, coming out with the same answer.

Enter special relativity

It was this belief that Einstein challenged when in 1905 he proposed his special theory of relativity. It would take us too far away along a rather technical route to describe all the details of this theory. Let us look at the motivation which led Einstein to it.

In Chapter 1, we referred to the work of James Clerk Maxwell on electromagnetic waves. While examining the basic equations that Maxwell had obtained for his theory, Einstein noticed that they implied a new kind of invariance for certain four-dimensional space and time combinations, an invariance that was somewhat similar to and yet somewhat different from the invariance of spatial distance that we have pointed out earlier.

We illustrate this new invariance with the help of Figures 5.7(a), (b). Let us consider two observers O_1 and O_2 in relative motion. Suppose, as viewed by observer O_1, observer O_2 is moving eastwards. Then observer O_2 will see observer O_1 moving *westwards* with the same speed. Both set their watches to zero time when they pass each other, at which time the distance between them is also zero.

The question now is, what will these observers find when they compare their metre rods and clocks? Let us take the metre rods first.

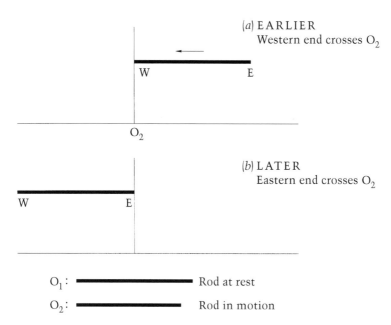

Figure 5.7: The observer O_2 sees the metre rod which is stationary in the frame of observer O_1 as slightly shrunk. In (a) the western end of the rod meets O_2 while in (b) the eastern end passes O_2 at a later time.

Suppose we have a metre rod lying west–east at rest in the frame of reference of observer O_1. Observer O_2 will pass its two ends at different times, crossing the western end first and the eastern end later. O_2 notes down these times and works out the difference between the two timings. Multiplying O_1's speed by this time interval gives O_2 the length of the rod.

The markings on the rod tell O_2 that it is a metre rod and so the result of the above measurement should be 'one metre'. Instead O_2 finds that the result is slightly less than a metre. In other words, *the rod appears slightly contracted when seen by a moving observer*.

A similar result obtains for time measurements. Suppose, as shown in Figure 5.8, O_2 passes two clocks at rest in O_1's rest frame, one after the other. Since O_2 is passing them at different times they will show different times. How will this time interval compare with the time interval recorded by the clock carried by O_2? Again, *the time interval recorded by the moving clock of O_2 will turn out to be slightly shorter than the time interval recorded by the stationary clocks of O_1*. Thus O_1 will think that O_2's clock is running slower.

What we have described above are *thought experiments*, but they do reflect how actual systems seem to behave in nature as well as in a terrestrial laboratory. Observations of fast-moving particles in cosmic ray showers have, for example, confirmed the slowing down of time, or *time dilatation*, for the particles called μ mesons. A typical μ meson, when at rest, decays in a time of about two microseconds. A fast-moving μ meson will, however, appear to last for a longer period, because in our frame of reference (we are like the observer O_1 in the above thought experiment), the clock determining the decay of the particle moves slower. Thus cosmic ray mesons are observed to last even as long as fifty times the above decay period.

The operative word 'fast' means that the particle is moving with a speed very close to the speed of light. The effects of length contraction and time dilatation are noticeable only when the relative motions involved are comparable to the speed of light. The typical motions encountered in daily life are much too small for these effects to be readily apparent. For example, if O_2 is travelling in an aircraft at a speed of 1 000 km per hour, the slowing down of clocks described above will be by a fraction as small as five parts in ten thousand billion.

Even though these effects are minuscule in our daily routine, they are counter-intuitive. We are so much used to the measurements of

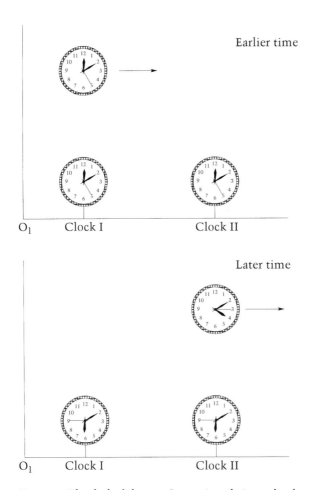

Figure 5.8: The clock of observer O_2, moving relative to the observer O_1, will appear to run slow compared to the two clocks stationary in O_1's frame.

spatial distances or temporal intervals being absolute that the idea of their being different for different observers appears very strange. That is why the special theory of relativity was initially violently resisted, even by philosophers and intellectuals in general. Several paradoxical situations were thought up to show that these ideas of space and time measurement were wrong. We will describe one such paradox shortly.

But to return to Maxwell's equations. The mathematician Hermann Minkowski showed that these apparently strange results of space and time measurement arise because of looking at them separately, rather than in a holistic way. Our example of city paths will illustrate what he meant.

Referring back to Figures 5.2 and 5.4, suppose in the two arrangements of paths shown there we were looking for the distance between A and B by confining ourselves to measurements along avenues only, ignoring the streets altogether. Then, as seen from these figures, the walk along an avenue is longer in the case of Figure 5.4 than for Figure 5.2. So we have to conclude that the 'distance' (as measured along avenues only) between A and B is different for the two cases.

This conclusion is evidently wrong, based as it is on our incomplete rule for measuring distances, ignoring the streets altogether. Had we taken the streets into consideration, drawn our right-angled triangles and used the Pythagoras theorem, then we would have discovered that the distance AB is independent of the system of streets and avenues chosen. It is invariant.

Minkowski's idea takes us one step further. It tells us that the true invariant distance between two points in space and time is not just the distance measured in space or just the interval measured in time: *it is a combination of both*. Again a theorem like that of Pythagoras is called into operation, but in a slightly different fashion. The new rule is not difficult to understand.

In Figure 5.9 we have a *spacetime diagram*. We have indicated space by the horizontal axis and time by the vertical axis. Actually, the space itself has three dimensions but we cannot draw all three on a plane piece of paper. But this shortcoming does not come in the way of our

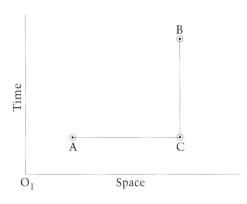

Figure 5.9: A spacetime diagram in which three events are shown. Events A and C have the same temporal epoch while B and C have the same spatial location as measured by observer O_1.

understanding the new rule for measuring the distance between two points A and B in spacetime. Note that when we plot a point A in this diagram, we are describing the location of an *event* in space *as well as* its epoch in time. The same applies to B. So we measure the spatial separation between the two events as well as the temporal interval separating them.

Now, in analogy with our street–avenue system, we imagine the diagram in Figure 5.9 criss-crossed with horizontal and vertical lines, the former being lines of constant temporal epoch and the latter lines of constant spatial location. A horizontal line through A will intersect a vertical line through B in a point C, just as we had in Figure 5.3. As before we have a triangle ABC with, apparently, a right angle at C. 'Apparently', because we have not really defined how to measure an angle between a time line and a space line. Indeed, the new rule that tells us how to measure the distance AB will be different from the more familiar Pythagoras theorem. The rule is as follows.

Multiply all time intervals by the speed of light so that they are now measured in distance units. AB squared is defined as equal to the *difference* of AC squared and CB squared. The 'separation' between any two events defined in this way is invariant with respect to the space-time frames used by all observers in uniform relative motion with respect to one another.

Let us go back to our two observers O_1 and O_2 in uniform relative motion. Suppose that Figure 5.9 is the spacetime diagram for observer O_1. How will the spacetime diagram of observer O_2 compare with it? Going by our analogy of the two path systems of Figures 5.2 and 5.4, we find that the lines of constant time and constant space for observer O_2 will be tilted with respect to those shown in Figure 5.9. Thus the two intervals AC and AB will be different from Figure 5.9. However, *if we use the above prescription for measuring the separation AB, we will get the same answer in both cases.*

Now, if we use this notion of combining space and time and define invariant separation in the above fashion, we will find that the Maxwellian equations of electromagnetic theory look the same for all such moving observers. That is, all observers in uniform relative motion with respect to one another will arrive at the same formal structure for these equations from their experiments. *Only* if we use the above modification of the Pythagorean rule will this symmetry come about. This was the motivation that led Einstein to this novel method of combining space and time into a single entity. Henceforth we shall refer to this combined entity as *spacetime*. It has four dimensions, one of time and three of space.

The speed of light

The above line of thinking gives the speed of light a very special status. For, a consequence of Maxwell's equations is that electromagnetic disturbances travel with the speed of light. This means that for all observers in uniform relative motion with respect to one another the speed of light will appear the same.

Once we accept the basic premise about the 'sameness' of the electromagnetic equations for all such observers, the above result seems natural. Yet, it also leads to some consequences that are against our intuition. A typical example from our normal experience will illustrate the difficulty.

Suppose you are travelling in a train at a speed of 100 kph. If another train approaches you at, say, 110 kph, it seems to come very fast, because its effective speed towards you is 100 + 110, that is, 210 kph. The train flashes past without your noticing the details of the carriage windows, people within etc. However, if you see the same train follow up and overtake yours, you will definitely see it in all details as it creeps up on your train and slowly overtakes it. The effective speed with which it overtakes your train is now only 10 kph, being the difference of the two speeds. Thus, in one case the train's speed seems very large and in the other very small. This example is typical in the sense that the apparent speed of the second train depends on whether it is approaching you or receding from you.

Now substitute light for the second train and you will realize the problem. For, we have just arrived above at the conclusion that *the speed of light will be the same whether you are approaching or receding from it*!

It was an important historical experiment that brought home this strange behaviour of light to the physicists, although they did not grasp its significance until after the advent of special relativity. The circumstances were briefly as follows. In the nineteenth century, there was a general belief that light waves require a medium in which to travel, a belief fostered by other familiar examples of wave motion such as water waves, which travel in water, and sound waves, which also require a medium, such as air, water etc. The expectation was that there is a tenuous medium called the *aether*, which is all pervasive and which is perturbed when a light wave crosses it. Can we detect its existence by, say, measuring the speed of the Earth relative to it?

Going by a reasoning similar to our example of the trains, scientists A.A. Michelson and E.W. Morley set up a very sensitive experiment to detect this speed. (See Figure 5.10 for a layout of this experiment.) Since the Earth rotates from west to east, a ray of light making a return trip in the east–west direction was expected to take a slightly longer time than one making a return trip of the same length in the north-south direction. Similarly, it can be shown that a boatman, rowing at a fixed speed with respect to the water surface, takes a shorter time to cross a river of

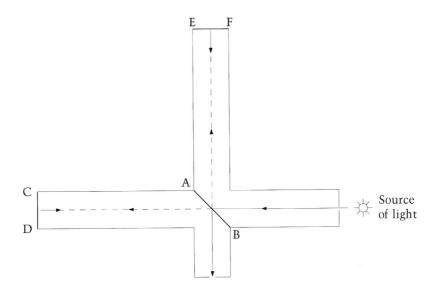

Figure 5.10: In the Michelson-Morley experiment, a ray of light (from a source on the right) falls on a partially reflecting and partially transmitting mirror *AB* in Michelson's apparatus. The reflected part goes up and is reflected back by the mirror *EF*. The transmitted part goes in the original direction and is reflected back by the mirror *CD*. The two parts recombine and are seen by the observer at the lower slit. In the extreme case where the crests of the two waves fall together the total light doubles, while in the opposite extreme the crest of one wave cancels the trough of another. In general, a series of dark and bright fringes is seen by the observer. The interference pattern of two waves depends on how far each wave has travelled as well as on the speed of light. Since the two arms of the interferometer were equal in length, the shifts in the interference fringes could be used to detect minute changes in the speed of light. Michelson and Morley used this technique to measure the expected difference in light travel time in the north–south and east–west directions. They failed to find any difference.

width *d* and then return than to proceed the same distance *d along* the river and then return. Despite several efforts to detect this minute difference none was found. The reason, as we now know from special relativity, is that the light speed will apear the same for all these journeys, regardless of which way the earth spins.

That light has a special status in relativity theory is also clear from Figure 5.9. Suppose that events *A* and *B* are such that they are connected by a light ray; that is, a light ray from *A* passes through *B*. Then, by our prescription, the lengths *AC* and *CB* are equal, and therefore the measure of *AB* is zero. And because it is invariant for all observers, all of them will see light move with the same speed, regardless of its direction.

Figure 5.11 describes this result through the concept of a *light cone*. If we send out a number of light signals in different directions from a point *A* in spacetime, they will all travel outwards on trajectories that lie on a cone, called the *future light cone* of *A*. Likewise light rays approaching *A* from all directions lie on the *past light cone* of A. Since the theory of relativity also leads to the conclusion that no material particle can travel with the speed of light, all such particles leaving the point *A* will have trajectories that will lie inside the future light cone of *A*. We normally expect that physical processes follow the principle of causality, that is, causes precede effects. Since no physical action can proceed faster than light, one can go further and say that *all causal effects* from *A* will lie inside or on the future light cone from *A*. That is, no physical effect of any kind can travel faster than light.

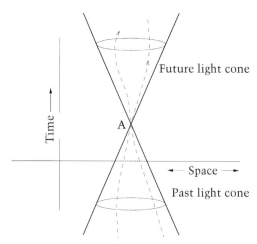

Figure 5.11: The future and past light cones from a general spacetime point *A* are shown. Any material particle sent out from *A* will have a trajectory *inside* the future light cone, as shown by the broken line.

The trajectories of material particles in the spacetime diagram are called *worldlines*. Light rays travel along the points of zero separation and are therefore called *null lines*.

This light speed limit was very difficult to understand at first, as the Newtonian ideas did not place any such limit. Nevertheless, by adopting special relativistic rules of motion and spacetime measurements, one could demonstrate that paradoxical situations arise if we permit faster-than-light transmission of information. The well-known limerick expresses one such situation:

> There was a girl named Miss Bright
> Who travelled faster than light.
> She departed one day in an Einsteinian way
> And came back yesterday night.

The clock paradox

There is one aspect of our observers in uniform relative motion that we have not yet clarified. According to the special theory of relativity, they form a special class of observers, called *inertial observers*. What are these inertial observers?

In formulating his ideas of motion in the seventeenth century, Isaac Newton had stated three laws. The first law of motion, which concerns us here, states that a material body continues to be in its state of rest or uniform motion provided no external force acts on it. So, we define our inertial observer as one on whom no external force acts. All such observers will continue to move with uniform velocities with respect to one another.

A paradox posed in the early days of special relativity brought out the role of the inertial observer clearly. Known as the *clock paradox* or the *twin paradox*, it generated considerable discussion amongst scientists and philosophers. It is worth mentioning briefly here.

Twins *A* and *B* set up an experiment. *A* stays at home while *B* sets off with a speed very close to (but, of course not equal to) the speed of light, on a journey into the cosmos lasting several days by his clock, during which he goes a long distance with the same speed, then slows down to a halt and reverses his speed and comes back. Both on his onward and return journey, he has mostly been travelling with a speed very close to the speed of light as measured by his twin *A*, and therefore, his watch runs very slow compared to the watch of A. So on his return B finds that A has aged (as indeed all people on the Earth have) by several years.

Where is the paradox in all this? Well, look at the whole experiment from B's point of view. He sees his twin A take off with a very high speed and return with high speed. So, by the same argument shouldn't B have aged compared to A? When they both get together after the completion of the experiment, they should be able to decide this fact one way or the other. So, which twin returns younger than the other and why?

On the face of it, it may apear that the experiences of A and B are symmetrical. In fact they are not: twin A satisfies our criterion of an inertial observer while twin B does not. B first accelerates in order to achieve his high velocity, later decelerates until his velocity is zero, then speeds up in the return direction until his velocity is the opposite of its original value and finally decelerates to come to rest on the Earth. *Thus B is not an inertial observer.* Figure 5.12 shows the worldlines of both the twins together to demonstrate this difference.

To see what happens to twin B and also to look through his eyes, one must include these changes of velocity. We find that whichever way we do the calculation the answer is the same: twin A is older than twin B.

SPACE, TIME AND GRAVITATION

Ten years after he proposed special relativity, Einstein came up with an even more profound theoretical exercise, which became famous as the

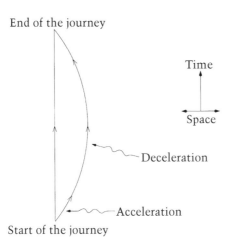

End of the journey

Time

Space

Deceleration

Acceleration

Start of the journey

Figure 5.12: The twin paradox. The worldline of twin A is straight and vertical while the worldline of twin B is curved, showing that he is *not* an inertial observer.

general theory of relativity. In this theory he addressed some outstanding issues relating to gravitation.

The law of gravitation which had been propounded by Isaac Newton in the seventeenth century had the hallmarks of a great theory: it was simple in its statement yet wide in its applications. It had proved successful in explaining phenomena on the terrestrial scale and within the solar system as well as for distributions of stars. Yet, by the first decade of this century it was becoming clear that the Newtonian theory may at best be an approximation to a wider theory of gravitation, that it had lacunae which needed sorting out.

Two such problems were as follows. The first one placed the Newtonian theory in direct conflict with the special theory of relativity. Recall that the special theory places a speed limit on the transmission of any physical effect from one point to another in space, the limit being the speed of light. Newton's law of gravitation does not respect such a speed limit. According to it the effect of gravitational attraction propagates across space *instantaneously*. Hermann Bondi has given an example of this conflict with relativity through the following thought experiment.

Imagine a situation whereby the Sun is suddenly made to vanish by magic. When on Earth will we notice the effect of this catastrophe? As sunlight travels to the Earth in about 500 seconds, we will notice the absence of the Sun from the sky 500 seconds after the event. However, if Newtonian gravity were right, we would notice the lack of gravitational pull towards the Sun immediately after the event. As shown in Figure 5.13, the Earth would cease to move in its elliptical orbit and would take off in a tangential direction. Thus we would be aware of the change in the Earth's motion even when the Sun is still visible to us.

Of course, in reality the Sun cannot vanish all of a sudden. The law of conservation of matter and energy tells us that something cannot just disappear out of existence. However, one may restate the situation by saying that the Sun may undergo a change in shape, or may collide with a passing star. Whatever happens, its gravitational impact would be felt by the Earth 500 seconds before the actual sighting of the event. A more consistent theory should make the gravitational effects travel with the speed of light so that both the visual and gravitational effects are noticed at the same time.

The second problem arises in the very definition of inertial observers. These are observers who never feel any force. *But, can such observers exist at all?* If we look at the situation more closely, we find that there is

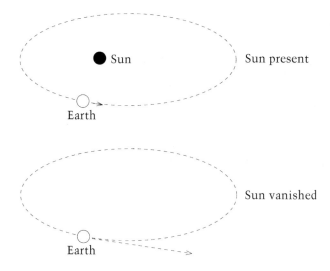

Figure 5.13: If the Sun were to vanish suddenly, the Earth would take off in a direction tangential to its orbit.

one force that is always present everywhere. Even if it is minuscule, it cannot be eliminated or shielded against. That is the force of gravity. The real universe is not empty anywhere: it contains matter and radiation and these will exert a gravitational force on anything anywhere. Thus inertial observers, so basic to relativity, do not actually exist.

At this stage we may recall pictures like the floating astronaut inside the Space Shuttle (see Figure 5.14). Is it a case of no gravity? No. It is a case of *micro-gravity*, which means that a tiny gravitational force is still present. For example, the walls of the shuttle exert a small but non-zero gravitational force on the astronaut. In fact, one can show that the force of gravity cannot be eliminated altogether under any circumstances. Contrast this behaviour of gravity with electricity or magnetism. One can devise a chamber in which there is no electric or magnetic force felt; the walls of such a chamber act as shields which prevent any outside forces from percolating in. Such shielding, though, is not possible for gravity. Gravity permeates everything and is a permanent feature of space and time.

I have emphasized this aspect of gravity because it served as the key property that made Einstein take the bold step of identifying it with the geometry of spacetime.

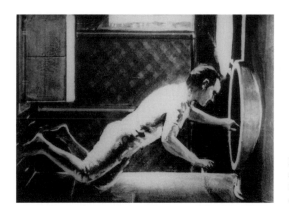

Figure 5.14: A floating astronaut in the Space Shuttle (picture by courtesy of NASA).

Gravity as we explained above, represents a permanent feature of spacetime; but so also does geometry, which describes how the lengths, intervals and angles in space and time are measured and what theorems hold for the various figures drawn in spacetime. However, just saying 'gravity is geometry' will not get us very far. To give some quantitative structure to this identification, one proceeds as follows.

Non-Euclidean geometries

The Oxford dictionary describes geometry as the 'science of properties and relations of magnitudes in space'. The first systematic account of geometry was given by Euclid around 300 BC. Euclid's geometry is taught to this day and it is this geometry that is most commonly used in everyday life, for example in the construction of buildings, bridges, tunnels etc.

Euclidean geometry – as any other branch of mathematics – starts with a small number of axioms or postulates. These are statements whose truth is taken for granted, and the entire subject rests on them just as a building rests on its foundations. If the postulates are changed the subject which is based on them also changes.

It took mathematicians many centuries to realize that Euclid's postulates were not sacrosanct: they could be changed and geometries other than Euclidean could be formulated in a logically self-consistent manner. The work of Lobatchevsky (1793–1856), Bolyai (1802–60), Gauss (1777–1855) and Riemann (1826–66) led to many non-

Euclidean geometries. How they differ from Euclidean geometry can be appreciated from a few examples.

First let us consider what is meant by a 'straight line'. In Figure 5.15(a) we see a line (drawn on a plane) which is not straight. At each point of the curve if we draw the tangent the direction of the latter changes as the point moves along the curve. However, in the case of a *straight* line this direction does not change. In Figure 5.15(b) we see another way of deciding which curve is straight. Of the lines connecting the two points A and B only the broken line is straight, it being the line of shortest length between A and B. If a rubber band is stretched between A and B it will tend to assume the shortest length and will lie along the broken line.

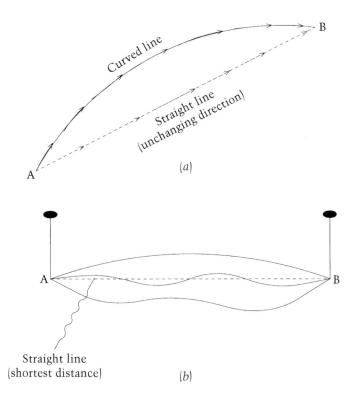

Figure 5.15: A straight line can be defined in two ways, (a) as the line whose direction always stays the same as one moves along it, and (b) as the line of shortest distance between two points.

To us, accustomed to drawing lines on plane paper, these properties of straight lines are intuitively acceptable. We can also accept Euclid's parallel postulate, which tells us that given a straight line *l* and a point *P* outside it, we can draw one and only one straight line through *P* parallel to *l*. Indeed, so plausible did this result seem that many mathematicians tried to prove it as a theorem from the rest of Euclid's postulates. But to no avail.

Eventually, it dawned on them that it is a postulate which has to be added to get the usual theorems of Euclid's geometry. Furthermore, it is not necessary to retain this postulate for a self-consistent geometry. For instance, we can construct a geometry on the assumption that through *P no* line can be drawn parallel to *l*. We can also go the other way: we can assume that through *P more than one* line can be drawn parallel to *l*. These alternatives become acceptable if we discard our intuitive notion of *flat space* such as the two-dimensional space of the plane paper described before.

Imagine, instead, the curved two-dimensional surface of a sphere. What would the geometry be like if lines are constrained to be on the sphere? As shown in Figure 5.16, we can draw a 'straight line' between any two points *A* and *B* on the sphere by stretching a rubber band between them. This line is in fact the arc of the great circle passing through *A* and *B*. (A great circle is the circle cut out on the spherical surface by a plane passing through the centre of the sphere).

Now any two great circles intersect and so all 'straight' lines on the spherical surface intersect. This result brings us into conflict with the familiar concept of parallel lines. Two straight lines drawn on a plane surface are considered parallel if they never meet even if extended indefinitely in either direction. Clearly this concept does not hold on the surface of a sphere. In other words, there are no parallel lines and we have here an example of the first type of violation of Euclid's parallel postulate. Euclid's proofs which make use of parallel lines naturally fall apart in the geometry of the spherical surface. For example, any triangle *ABC* will have its three angles adding up not to a sum of 180° as Euclid would have it, but to a sum greater than 180°. The triangle shown in Figure 5.16 has $\hat{A} + \hat{B} + \hat{C} = 270°$.

Curved surfaces of this type are called surfaces of *positive curvature*. If we had instead elected to modify the parallel postulate in the second of the two ways described before, we would have arrived at a geometry which applies to surfaces of *negative curvature*. A saddle, or the surface

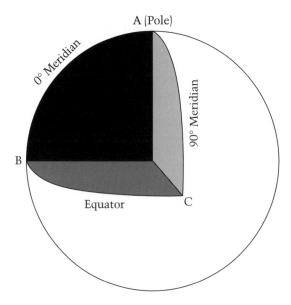

Figure 5.16: A triangle *ABC* on the surface of a sphere has its three angles greater than two right angles. For this particular spherical triangle, each of its three angles is a right angle. Note that 'straight' lines can be drawn by stretching rubber bands between *A*, *B* and *C*.

of a jar near its lip are examples of surfaces of negative curvature. For a triangle *ABC* on such a surface we have $\hat{A} + \hat{B} + \hat{C}$ less than $180°$.

There is an easy experiment to decide whether a surface has a zero, positive or negative curvature. Take a piece of paper and try to cover parts of the surface with it. If the paper lies plush on the surface, it has zero curvature, that is, it is flat. If, on the other hand, the paper gets folds and wrinkles in your attempts to cover the surface exactly, then the surface has a positive curvature. The third possibility is that if the paper gets torn in the covering process, then the surface has a negative curvature. Try this experiment on a table top, a sphere and a saddle.

What has all this to do with gravity? The concepts of flat and curved space can be extended to spaces of higher dimensions. For example, the geometry of the three dimensions of space and one dimension of time in which Einstein's special theory of relativity applies is the geometry of *flat* space. Because of the ever-present gravity, this flat spacetime geometry is, according to Einstein, an idealization. The geometry of space and time must in reality be of a curved non-Euclidean type. This important conclusion of Einstein is often stated in the form 'spacetime is curved'.

The effect of matter on spacetime geometry

From the above examples of geometries of space, we now have to generalize to the geometries of spacetime. It is best done with the help of an example of a familiar event: a ball being tossed vertically upwards.

Figure 5.17 is a spacetime diagram showing the worldline of the ball. On the horizontal axis we plot the height from ground level, while on the vertical axis we plot the time elapsed since the ball was tossed up.

In brief, this is what happens. The ball has been tossed up with a certain speed, say 12 metres per second. It loses its speed, however, as it climbs up and finally comes momentarily to rest at a height of 7.5 metres. Then it begins to drop. As it falls further, its downward speed increases, reaching the final value of 12 metres per second at the level of the initial throw. In the figure, the worldline appears as a curve which mathematicians call a *parabola*. The curve starts with a slope which depends on the initial speed of throw and it gradually steepens up as the speed declines. At the maximum height reached by the ball, the curve is momentarily vertical and thereafter it begins to move towards the time axis (corresponding to the downward motion of the ball).

How do we interpret this motion according to Newton's laws of motion and gravitation? The ball behaves in this fashion because the

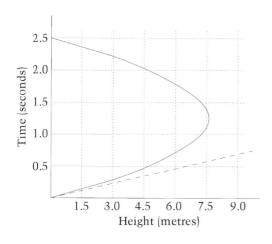

Figure 5.17: The worldlines of a ball tossed up vertically in reality (continuous curved line) and in the case of no gravity (broken straight line).

Earth's gravity attracts the ball downwards, causing a deceleration of its speed. In fact, had there been no such force of gravity, the ball would have continued to travel with the same speed upwards. (Newton's first law of motion tells us that in the absence of any external force a body maintains its speed and direction.) The broken straight line shows this hypothetical behaviour.

This broken line is an example of a straight line in the four-dimensional spacetime. That is, if we have the spacetime described for special relativity (see Figure 5.9), this line would qualify for being called 'straight'.

As discussed above in the case a curved surface such as that of a sphere, we can identify 'straight lines' in a curved spacetime by an extension of the technique of stretching a rubber band between two points. The technique is, however, too mathematical to be described here.

Now let us consider the two lines of Figure 5.17 on Einstein's terms. Einstein would argue that the hypothetical question of how the ball would have moved in a 'no gravity' situation has no bearing on the actual situation. For such a state could not be realized – gravity being indestructible. The only worldline that has any reality is the curved one. This is the one we have to understand and interpret in as simple a way as possible.

Einstein's argument would thus be that *the solid curve of Figure 5.17 is the real straight line*. It corresponds to uniform motion under no forces. 'No forces', because gravity as a force has been replaced by a non-Euclidean spacetime.

Now, such a statement appears manifestly wrong: surely, the line is curved and cannot be called a straight line. Moreover, the speed of the ball along it is also not the same everywhere.

However, which line is straight and which line is not straight depends on the rules of the particular geometry. In Figure 5.17 we were tacitly using Euclid's geometry to settle the issue. *Einstein would argue that the geometry is not Euclidean: the presence of Earth's gravity makes the geometry of spacetime non-Euclidean.* This is why, with the changed rules of geometry, the continuous line of Figure 5.17 qualifies for the adjective 'straight'. The same comment applies to the apparent change of speed. If the rules of the non-Euclidean geometry are applied to the continuous trajectory, we will discover that its speed, interpreted in four dimensions, is constant in magnitude and direction.

Perhaps an analogy with the maps in geography atlases will help. These maps often show latitude lines as straight. However, on the curved surface of the Earth, these lines are *not* straight. That they are not lines of shortest distance can be easily verified by stretching a rubber band between two points on the globe at the same latitude.

So Einstein's basic strategy towards a gravitational theory was the following. Any distribution of matter and energy in space necessarily makes the geometry of spacetime non-Euclidean. Given such a geometry, the worldlines of bodies moving in it are straight lines, that is, trajectories computed on the assumption that the body is moving in the spacetime with uniform speed and in an unchanging direction. Such straight lines are given the more technical name 'geodesic'.

In this way Einstein would argue that a body on which no other force than gravity acts will move along a geodesic, *computed according to the rules of the prevailing geometry*. Einstein gave a set of equations to determine the prevailing geometry if the information about the distribution of matter and energy in the region were known.

This is essentially what the general theory of relativity is all about.

Solar system tests

The general theory of relativity was first applied in 1916 to the problem of the motion of planets in the solar system. The method of solving the problem used by Karl Schwarzschild (Figure 5.18) was as outlined above. That is, assume that the space contains a spherical ball with mass that of the Sun and then work out with the help of Einstein's equations what the geometry of spacetime will be like around it.

Fortunately the problem can be exactly solved, despite the rather complex character of Einstein's equations. The Schwarzschild solution is considered very basic to relativity and has been used in many different contexts, including the solar system. The geometry is, of course, quite different from Euclid's and therefore the geodesics in this geometry are not the same as Euclidean straight lines.

Using the philosophy of the ball-tossing example, we can compute these geodesics in Schwarzschild's geometry to work out how planets move around the Sun. For, planetary worldlines are the geodesics of Schwarzschild's geometry. A typical worldline is described in Figure 5.19. If we project it onto the space part of the spacetime diagram, we get planetary orbit. *For all practical purposes, this is almost the same orbit that we would get if we used Newton's laws of motion and gravitation.*

Figure 5.18:
Karl Schwarzschild.

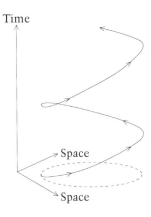

Time

Space

Space

Figure 5.19: The worldline of a planet spirals up in such a way that its projection in space will describe the elliptical orbit of the planet.

Almost, that is, but not exactly! There are small differences that are imperceptible by observation except for the planet Mercury. We referred to this small effect in the last chapter (see Figure 4.18). The orbit of Mercury slowly precesses so that the direction from the Sun to the closest point of the orbit (called the *perihelion*) rotates in space at a small angle of 43 seconds of arc per century. As we mentioned there, this general relativistic effect cleared up the long-standing mystery surrounding Mercury's anomalous behaviour, observed since the middle of the last century.

There have been other solar sytem tests too, of which one involving the 'bending of light' has had many consequences. It was this test, and the dramatic way its results were announced, that established the general theory of relativity as something very revolutionary in the public mind. But more of this story in the following chapter.

A more recent test, to which reference was also made in the previous chapter, became possible with space technology. It involves the delay in the echo of a radar signal when it passes close to the Sun's surface.

All these tests go towards demonstrating that the simple Euclidean geometry, which we have taken for granted on the basis of its use on Earth, may not exactly describe reality. Likewise, Newton's laws of motion and gravitation, which have served us so well, may not be entirely correct. A comparison of the actual universe with the predictions of Newton and of Einstein results in those of the latter being found closer to reality. And so we should adopt Einstein's way of looking at the phenomenon of gravity.

A figurative way of expressing this conclusion is to say that we live in a curved spacetime. Because the gravitational effects on the geometry of spacetime are relatively very modest in our vicinity, we could manage quite well with Newton's laws and Euclid's geometry. However, the cosmos has other locations where gravitational effects are very strong, with the result that the spacetime may behave very peculiarly when judged by our Euclidean standards. We will now describe such remarkable examples.

GRAVITATIONAL COLLAPSE

Intuition tells us that the effect of gravity will be large wherever there is a large concentration of matter and energy. How do we arrive at such large concentrations? Before we answer this question let us examine a way of telling whether the force of gravity in a region is strong or weak. This is provided by the notion of the escape speed.

Escape speed

Let us go back to our example of the ball tossed up vertically. We noted that if thrown with an initial speed of 12 metres per second, it will rise to a height of 7.5 metres. How high will it rise if we double the initial speed? Calculation tells us that it will rise four times higher.

This calculation takes into account the force with which the Earth attracts a body through Newton's law of gravitation. Although we have been arguing that general relativity is the superior theory, the answers provided by Newton's laws are good enough for this discussion.

We can push our argument further. Can we throw the ball with such a large speed that it never returns? At first the answer seems negative. For, by a naive extrapolation of the above argument it would appear that no matter how high we throw the ball it would return. But there is a catch. As the ball goes higher, as it moves away from the Earth, the gravitational force loses its intensity. At a height equal to the radius of the Earth, some 6400 km, the force drops to one quarter of its value at the surface of the Earth. And it goes on dropping at greater heights. It is therefore possible to send up the ball at a limiting speed at which it just keeps going farther and farther, never to return. This speed, appropriately called the *escape speed*, works out to be approximately 11.2 km per second, that is about 40 000 km per hour.

This minimum speed limit is way beyond what the fastest aircraft have achieved on Earth. But it is not beyond the capacity of our powerful rockets. Thanks to our space technology, it has been possible to escape from the Earth, and it has even been possible to launch spacecraft like Voyager I and II that have left the confines of our solar system.

The escape speed of a region therefore tells us how strong the local force of gravity is in that region. On the one hand, take the Moon. The escape speed from the Moon's surface is a mere 2.4 km per second. This is why it was relatively easy for the lunar astronauts to make the return trip from the Moon. They could lift off their small spaceship with built-in rockets rather than requiring an enormous rocket establishment like the one at Cape Kennedy.

On the other hand, on Jupiter the escape speed is much higher, about 60.8 km per second, while at the Sun it is more than ten times higher, equalling some 640 km per second. And on the surface of a neutron star it may be a whopping 160 000 km per second! This gives us an indication that here on Earth we live in a very modest gravitational environment. In the cosmic setting, though, we may encounter strong gravity the like of which we cannot imagine here.

Having got some idea of how the strength of gravity can be estimated, let us see how regions of strong gravity tend to be built up in the cosmic environment.

The build-up of highly collapsed objects

Gravity, whether you look at it through the eyes of Newton or of Einstein, is a strange kind of interaction. When Newton was asked if he had probed deeper into the origin of this law of gravitation, as to why such a law operated in nature, he remarked: '*Non fingo hypotheses* (I do not frame hypotheses)'. His approach was essentially empirical, to observe natural phenomena, look for a pattern and see if it followed from a simple but general law. Einstein did go further in seeing a connection between gravity and spacetime geometry. But here too we have not yet made further progress in understanding the basic cause of this link. In particular, an understanding of gravitation at the microscopic level of matter, using the rules of quantum theory, still eludes us.

But, based on what we know so far, it is possible to extrapolate into the unknown, and this is what we will do in our description of a massive object that is striving unsuccessfully to maintain an equilibrium despite its inward pull of gravity. For, as we will soon find, such an object will take us to a state of strong gravity.

Figure 5.20 shows such a massive spherical object. Symbolically we denote by A, B, C etc. its component parts, all of which are attracting one another. The net effect is to make the object itself shrink, unless some outward force prevents this from happening. In Chapter 2 we saw that stars have to face this problem continually and that they are able to maintain their equilibria through internal pressure, *provided they generate this through nuclear reactions at the centre.* When they run out of nuclear fuel, they may still have another chance, through the degeneracy pressures arising from heavy compression of their matter. Such stars then survive as white dwarfs or neutron stars. But in each case there is a limit on their masses. For a white dwarf, the mass must not exceed forty per cent above the Sun's mass, while the limit for a neutron

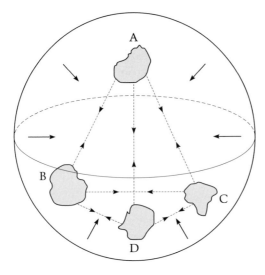

Figure 5.20: The different parts of a body, $A, B, C, \ldots$ attract one another gravitationally, with the result that the entire body tends to contract.

star is somewhat higher, but barely touches two solar masses. What happens to a star that finds itself with a larger mass than these limits when its nuclear fuel is exhausted?

In Chapter 3 we saw the possibility that a massive star would explode as a supernova, leaving behind a core. We could frame our above question for the core instead.

Here we recall the controversy between Eddington and Chandrasekhar recounted in Chapter 2. Eddington had refused to believe Chandrasekhar's result for the maximum permissible mass of a white dwarf, because he was worried about what would happen to those stars whose masses are well above these limits. We now address the situation that these stars will go on contracting without any significant internal pressure to oppose gravity, just as Eddington had feared.

A solution to the problem of contracting balls of dust was first discussed by an Indian relativist, B. Datt, in 1938. The term 'dust' here means matter without any internal pressure. Since we are considering situations wherein no significant internal pressures are available to resist gravitational contraction, the dust assumption is not far from reality. Here we notice a rather unusual behaviour of gravitation, which is not found in other known forces.

Figure 5.21 illustrates two different situations. In (a) we see two balls connected by a spring that has been stretched beyond its natural length.

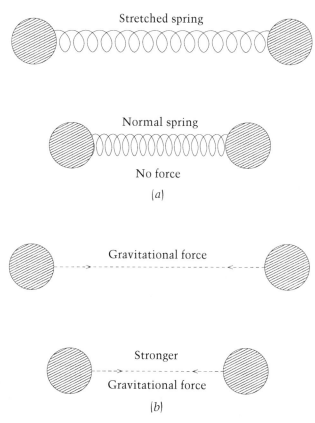

Figure 5.21: The contrasting behaviours of the elastic force and the force of gravitation are illustrated above (see the text for details).

The elasticity of the spring will make it contract and so the two balls will be attracted towards each other. If we let them slowly approach each other, the force of attraction declines and disappears altogether when the spring attains its natural length. In (b) we see two balls attracted by their mutual gravitation. As they move closer, however, the force of attraction does not diminish; rather it grows.

A spring-like force diminishes in strength if motions take place as required by the force. By contrast gravitation grows if motions follow its dictates. For these reasons Hermann Bondi has likened gravitation to a dictator who demands more if his earlier demands are acceded to.

Thus Eddington's apprehensions were well founded. As Datt's solution showed, stars with no adequate pressures to resist contraction will find themselves contracting more and more rapidly. Figure 5.22 illustrates how a dust ball contracts if it starts from rest. Notice that the rate of contraction is slow at first but starts growing rapidly until the contraction is catastrophic. This is the reason why scientists use the words 'gravitational collapse' to describe the situation.

In Figure 5.22 we have taken the dust ball to have the mass of the Sun, just to fix ideas. Notice that the entire ball shrinks to a point in a matter of 29 minutes. However, two cautionary remarks are needed while looking at Figure 5.22. First, the Sun itself will never have this fate. Having a mass less than the Chandrasekhar limit, it will settle down to the state of a white dwarf star. Our second remark is more basic. Recall that the theory of relativity does not have any absolute time. So what time are we using in plotting the figure here? By whose clock does the dust ball take 29 minutes to shrink? We shall clarify this point in the following section.

TIME DILATATION DUE TO GRAVITATION

According to Einstein, the effect of gravitation is felt not only in space but also in time. This becomes apparent in the discrepant rates at which clocks run in different regions. To be more specific let us consider two

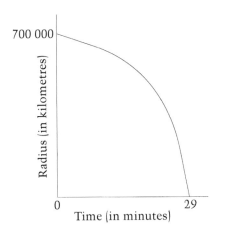

Figure 5.22: The contraction of a ball of dust, of solar mass. The contraction is slow to begin with but grows rapidly until the whole sphere shrinks to a point in about 29 minutes.

observers *A* and *B* (Figure 5.23) in different regions of space. Near *A* there is very weak gravitational influence – so that the spacetime geometry is almost Euclidean. Near *B* the gravitational influence is strong. Let us also assume that there is no change in the situation with time – that is, the situation is static. Suppose that *A* and *B* communicate with each other with light rays and that they decide to use atomic clocks in their respective regions as time-keeping devices. Now, from our everyday experience we would expect that if observer *A* sends signals every second by this clock, observer *B* will receive them every second, and vice versa. But this is not what happens. To observer *B* the signals from *A* appear to come at *shorter* intervals than a second and, conversely, to observer *A* signals from *B* come at longer intervals than a second.

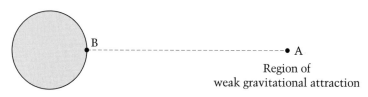

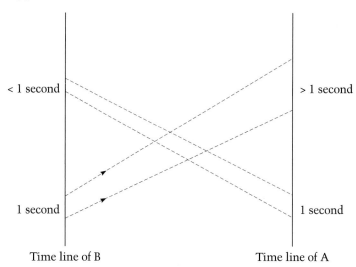

Figure 5.23: The time scales of two observers, one of whom is in a strong gravitational field and the other of whom is not, are different, as shown by the exchange of light signals.

Such a situation can arise in astronomy. We may identify the region near A with the Earth and the region near B with the neighbourhood of a massive compact object. To observers on the Earth the clocks on the massive object will appear to go slower. In practice, of course, we do not see the clocks going fast or slow. Instead, we see changes in the frequencies of spectral lines, because these reflect the time changes in the atomic systems at the source. Thus, in the above example, the frequency of light from the compact massive object will appear to be reduced, and the wavelength will appear correspondingly increased. Since the visible part of the spectrum ranges from violet at the short waves to red at the long waves, an increase in all wavelengths will cause the entire spectrum to shift towards the red end. This phenomenon is called *red-shift* and, as it is happening because of gravity, it is further qualified as *gravitational redshift*. We may therefore say that in the story at the beginning of this chapter, Brahma lived in a place of high redshift that is, large time dilatation!

We may look at this phenomenon from the microscopic angle too. Light is also considered as a swarm of particles called *photons*. The energy of a photon, as we saw in Chapter 2, is proportional to its frequency. With gravitational redshift, the frequency diminishes and so therefore the photon loses energy. This happens because the photon has had to spend energy to escape from the strong gravitational influence of the large mass.

Suppose we have a spherical body with uniform density undergoing contraction. We consider a typical particle B on the surface of the body, and study its motion inwards. By way of comparison we have an external observer A located far away from the body, practically out of the body's gravitational influence. As the body contracts, the gravitational strength in its neighbourhood increases, and the gravitational redshift effect begins to assume significance. Suppose A and B are in communication with each other in the way discussed above. The situation now is different in one sense, however, because while A is at rest B is moving inwards away from A. This leads to dramatic consequences.

As in the static case, B's clock appears to A to be slow. This effect now arises from two causes: first, the gravitational redshift and, second, the Doppler effect, because B is receding from A. The Doppler effect applies to wave motion in general. We notice it readily in the case of sound waves. The whistle of an approaching railway engine appears high pitched and it drops to a lower pitch once the engine passes us and begins to recede. Applied to light, this means that the frequency of a receding source drops and its wavelength increases. So again we have redshift.

The gravitational redshift and the Doppler redshift therefore add up for signals coming from B to A. In the case of observer B, however, the situation is different. The Doppler effect tends to lower the frequency of signals from A while the gravitational effect tends to increase it. Thus, as seen by B, A's clock appears to be running either slow or fast – depending on whether the gravitational effect is less or more important than the Doppler effect. It is clear, therefore, that the time scales of A and B are different. Let us continue to study the situation both from the point of view of B and of A.

According to B there is a continuous collapse towards the state of infinite density. The interesting result is that with B's time scale the contraction rate of the body follows exactly the same rule as in the Newtonian case. Thus, measured on B's clock, a star of the mass and radius of the Sun (but of pressureless and uniformly dense matter) will collapse to zero radius in 29 minutes. This answers the question raised at the end of the previous section. However, the similarity with the Newtonian case ends here. More drastic consequences are in store for B. As the matter in the collapsing object gets more and more dense, the geometrical properties of spacetime become more and more peculiar (that is, non-Euclidean), until finally a state of infinite density is reached. At that stage all geometrical description breaks down, since such a description involves mathematical operations with zero and infinity, operations which cannot be properly defined. This is a state of 'singularity'. It is very similar to the singularity of the big bang universe, which we shall encounter in Chapter 7, except that there the universe *explodes* from an infinitely dense state while here the body *implodes* to an infinitely dense state.

What does A see during this period? Does he see B falling into the singularity? The answer is *no*, and the reason is as follows. At the beginning, B's signals arrive at A separated by intervals of, say, slightly more than one second. These intervals get progressively longer (see Figure 5.24) as B falls further inwards – until a critical stage comes when B approaches a barrier known as the 'Schwarzschild barrier'. When B reaches this barrier his signals no longer reach A, no matter how long A waits. All information about B at that time and after he crosses the Schwarzschild barrier is inaccessible to A. It must be emphasized that at the time of crossing this barrier, B does not notice

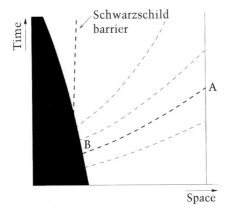

Figure 5.24: This spacetime diagram shows how the clock of B progressively slows down as observed by A, compared with the local clock at A. The light signals from B arrive at A at progressively longer intervals.

anything peculiar with the spacetime geometry at all. By B's clock everything goes on smoothly, until the singular state is reached.

We should emphasize that the Schwarzschild barrier acts only one way: it prevents signals from within getting out. B, however, continues to receive A's signals even after crossing the barrier, right up to the end.

BLACK HOLES

The Schwarzschild radius (that is, the radius of the spherical Schwarzschild barrier) for a body of mass M is given by a simple rule, $2GM/c^2$, where G is the gravitational constant and c is the speed of light. For the Sun the Schwarzschild radius is a mere 3 km. This is very much less than the Sun's actual 'radius', which is about 700 000 km. Only if the Sun shrinks from its present size to a radius of the order of 3 km will it become invisible to us.

Objects which are close in size to their Schwarzschild radius are almost invisible, because light from them is highly redshifted and has lost most of its energy. Such objects are called *black holes*.[8] By definition, they cannot be 'seen'; but they can be detected through their gravitational influence. For instance, if the Sun were to become a black hole, it would cease to be visible but it would continue to attract

[8]Mathematically speaking an object will be strictly a black hole when its radius *equals* the Schwarzschild radius. However, as we have just discussed, to external (*A*-type) observers like us, no object will ever be seen attaining that state.

the Earth. So the Earth would go round in an elliptic orbit with no apparent source!

The detection of black holes in the universe forms one of the most intriguing searches of astronomy. A black hole represents the ultimate negation of the conventional truth 'seeing is believing'. Since it cannot be seen its existence can only be inferred indirectly. We will discuss this issue shortly.

Perhaps the most relevant question about black holes is 'what sort of an object can become a black hole?' And here we encounter the difference between the gravitational theories of Newton and Einstein.

Suppose we consider an object a million times or more as massive as the Sun. How can we hold it in equilibrium? In the Sun, nuclear reactions generate internal pressures to withstand its self-gravitation. However, as we make an object more and more massive, these nuclear pressures tend to rise in proportion to the mass, whereas the self-gravitational force rises in proportion to the *square* of the mass. So, for an object with a mass equal to a million solar masses, nuclear reactions are unable to provide the necessary pressures to balance the gravitational force. Such an object would therefore collapse and become a black hole, unless, during the contraction, Nature stepped in to prevent this outcome, as Eddington felt would happen, and the object somehow got disrupted and broke up into smaller pieces.

So, can something prevent the gravitational collapse of a massive object? In Newtonian theory we could conceive of some 'new' agency with strong enough pressures to halt the collapse. In Einstein's theory the situation is different. Even if we succeed in inventing any such agency, its pressure must always be generated by energy. This energy, equivalent in relativity to mass, itself attracts and therefore helps the collapse. In the late 1960s, work by the theoreticians Roger Penrose and Stephen Hawking showed that, in general, unless we introduce new agencies with *negative energy*, collapse into a singularity is inevitable for most physical systems which have already contracted beyond a certain limit. Thus in the example discussed here, the collapse of B into a singularity *cannot be halted* once it has passed the Schwarzschild barrier.

Are singularities desirable in a physical theory? Normally physicists and mathematicians tend to frown on them and regard them as indicators of imperfections in the theory. Accordingly, it is possible to take the view that singularities in general relativity are undesirable and that we should look for 'better' theories. But there is the alternative view, formed probably because no better theories are in sight, that singularities take us to the frontiers of physics and that their existence is not a subject for debate in physics. We shall return to this issue towards the end of the book.

Does Cygnus X-1 house a black hole?

Perhaps the most interesting source among the X-ray binaries is Cygnus X-1, because very likely it contains a black hole. This X-ray source has been identified with a binary system of which the visible member is the supergiant star HDE 226868. The other member cannot be seen but its existence can be inferred from the observed motion of the visible companion. For, the visible member is seen to move in an elliptic orbit and this is not possible unless it has an invisible companion attracting it. The period of the binary system, determined from its optical properties, is 5.6 days.

In 1971 a faint radio source was detected in the vicinity of Cygnus X-1 by L. Braes and George Miley and by Campbell Wade and R. Hjellming. The variations of radio flux coincided with the variations of X-ray flux, leading to the conclusion that the X-ray source and the radio source are one and the same object. In fact this fortunate circumstance helped in the identification of the X-ray source with the optical binary system. For, unlike radio sources, which are comparatively rare, visible stars (including binaries) are very common, so that unless the position of the X-ray source is very accurately known it is a difficult problem to identify the precise optical object associated with it. Although the present X-ray telescopes can pinpoint a source within an angle of a few sconds of arc, in 1971 the source Cygnus X-1 could be pinpointed by the UHURU X-ray satellite only within an angular area of 4 square minutes of arc. With the far superior radio techniques available, the source could be located much more accurately within an angular area of 1 square second of arc.

It thus became possible to single out HDE 226868 and its invisible companion as the binary system generating the X-rays from Cygnus X-1. The more accurate X-ray detector in the Einstein Observatory later confirmed this identification.

The visible component of the binary system is a B-type star. In the spectral classification scheme of stars, such B-type stars are massive and luminous (see Chapter 2). From general information about the masses of these stars, the mass of HDE 226868 was estimated to be *at least* 20 solar masses. The period of the binary is 5.6 days. The velocity of the visible component in the radial direction can also be estimated and, using the Newtonian law of gravitation, the mass of the invisible companion can be estimated to be *at least* five solar masses. The reasons for the qualification 'at least' are that we are not necessarily located in the orbital plane of the binary system and that our estimate of the mass of the extended companion is a lower limit. Hence we cannot estimate the mass of the compact object exactly; we can only estimate the limit that the mass must exceed.

Even this lower limit of five solar masses is, however, well above the limit of the mass of a neutron star discussed in Chapter 4. What can the compact object be, if it is not a neutron star? It also is known that the X-rays associated with the binary fluctuate rapidly in intensity. The time scale of the fluctuations, 0.001 s, can be translated into a distance scale on multiplication by the speed of light, which gives a distance of 300 km. From relativity theory we know that no physical disturbance can travel faster than light. Thus the physical effect of any large-scale change is expected to travel within the source at a speed less than light. Hence any coherent physical process which produces fluctuations as rapid as a thousandth of a second cannot extend over a region larger than 300 km in size.

Recalling from Chapter 4 how binaries evolve, we note that the invisible companion will suck in matter from the visible star and this matter will fall into the former after circling round it for a while (see Figure 5.25). This spiralling matter forms a disc, called the *accretion disc*. The X-rays come from the heating of this disc. Now from what we have just seen, the accretion disc has to be as small as 300 km in size if it is to generate such rapid fluctuations of X-ray intensity. The high value of the X-ray intensity also enables theoreticians to conclude that the emitting source must be a highly compact object.

Evidence of this kind makes Cygnus X-1 almost unique amongst X-ray binaries. Since no other type of compact star can fill the bill, the conclusion drawn is that the unseen member of this binary system is a *black hole*. If this conclusion stands, then X-ray astronomy can claim the credit for the first discovery of a black hole!

Supermassive black holes?

The following lines could very well have come from the pen of Conan Doyle if his supersleuth were chasing a cosmic mystery:

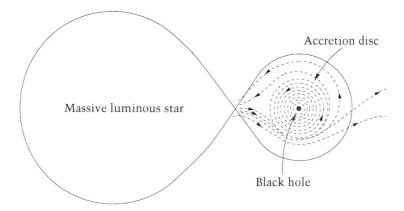

Accretion disc

Massive luminous star

Black hole

Figure 5.25: The scenario of Cygnus X-1 is depicted in this figure; for details the reader should refer to the text.

'I am convinced, my dear Watson, that a black hole is responsible for this violent act.'

'A black hole! Surely, Holmes, aren't you going too far?', I exclaimed incredulously. But my friend shook his head.

'Yes, a black hole and a supermassive one. How often have I said to you that when you have eliminated the impossible, whatever remains, however improbable, must be the truth?'

Modern theoreticians have found the improbable black hole an answer to the cosmic energy problem[9] on the grounds that no other less dramatic solution is available. Indeed, the 'black hole industry' really took off in astronomy after Cygnus X-1. Astrophysicists were confronted with other astronomical events wherein a black hole seemed to be the best way to account for observations. And unlike the black hole in Cygnus X-1, here one required a *supermassive* black hole, containing more matter than a billion Suns.

As just mentioned, the key issue in these very dramatic observations was to explain how *so much energy is ejected from such a limited space in such an explosive fashion.*

What were these events?

[9]Typically the problem is to find a mechanism for an enormous output of energy from a very compact source.

The first clue to them came shortly after the end of the Second World War, not from the age-old optical astronomy but from the newly emerging radio astronomy.

COSMIC RADIO SOURCES

In 1946, J.S. Hey, S.J. Parsons, and J.W. Phillips discovered radio waves coming from the direction of the Cygnus constellation. The techniques of radio measurement in 1946 were not accurate enough to pinpoint the location of the source. In 1951 F. Graham Smith at Cambridge was able to achieve sufficient accuracy in locating this source to enable optical astronomers to institute a search for a source of visible light in the same place. The radio source was called Cygnus A.

Walter Baade at the Mount Wilson and Palomar Observatories looked for and found an interesting object at the location of Cygnus A. Figure 5.26 shows the photograph of this object, now known as a *radio galaxy*. Baade in fact thought that the photograph showed two galaxies in collision. He believed that galaxies in collision would generate the kind of energy that was needed to power the radio source.

It is interesting to recall a bet which Baade made with Rudolf Minkowski, another leading astronomer at the Mount Wilson and Palomar Observatories. The bet arose when at the end of a seminar talk on Cygnus A, Minkowski made sceptical comments about the collision hypothesis proposed by Baade and Lyman Spitzer to account for the radio emission from Cygnus A. Baade was, however, confident enough to bet on his theory to the tune of one thousand dollars; but Minkowski talked him down to just a bottle of whisky! It was agreed by both sides that the evidence of emission lines in the spectrum of the gas in the source would be taken as confirmation of the idea that colliding galaxies were involved. A few months later this evidence was obtained and Minkowski conceded the bet. However, Baade later complained that Minkowski himself finished the whisky that he had given in settlement of the bet!

Subsequent events showed that Minkowski was justified in consuming the whisky, for later evidence confirmed his scepticism of the collision theory. Now it is realized that Cygnus A *does not* owe its radio emission to the collision of two galaxies. What goes on in Cygnus A is in fact characteristic of what goes on in the majority of radio sources

Figure 5.26: An optical photograph of the galaxy identified with the radio source Cygnus A (courtesy of Palomar Observatory, California Institute of Technology).

located outside our Galaxy; these have been discovered since 1951. The detailed evidence available in such cases points not to a collision but to an explosion in the central region of the radio source, an explosion which throws out electrically charged particles in opposite directions, as shown in Figure 5.27. These fast particles proceed a certain distance from the source and then radiate in the presence of the magnetic field in the region. What is the process that leads to the emission of fast particles from a radio source? Where does the source derive its tremendous power?

The energy problem

The first reasons for casting doubts on the colliding galaxies hypothesis for radio sources like Cygnus A were theoretical. In the late 1950s, Geoffrey Burbidge gave an elegant argument estimating the energy stored in a strong radio source like Cygnus A. Burbidge's calculations

Figure 5.27: Schematic diagram of a typical extragalactic radio source. The central region ejects fast particles, which radiate radio waves from two lobes located on opposite sides of the central source.

took into account all the observed properties of the radio waves coming from Cygnus A, including their intensity and spectrum, and made the hypothesis that the radiation was coming from fast-moving electrically charged particles accelerated by the magnetic field in the radio source.

From the available data Burbidge was able to estimate the *minimum* energy that must necessarily be present in the particles and in the magnetic field in order to sustain the observed radiation. Typical total energies in these two modes are comparable and work out to a staggering figure, even by astronomical standards. The energy required would far exceed the typical energy stored in a normal galaxy of stars like ours. By terrestrial standards it is equal to about ten thousand billion billion billion billion times the energy released in the explosion of a megaton H-bomb!

How much energy can a pair of colliding galaxies produce? The collision hypothesis depended on conversion of the gravitational energy of the colliding pair of galaxies into energy of radio waves. That is, in the process of collision, the gravitational energy would be used in accelerating charged particles to high speeds so that they could radiate. However, detailed calculations showed that this process could produce only a thousandth part of the requisite energy! Thus, dramatic though a collision of galaxies is expected to be, it is not powerful enough to sustain radio sources like Cygnus A.

Later observations in the early 1960s revealed the picture of Figure 5.27. The radio emission did not come from the central galaxy but from lobes located hundreds of thousands of light years away from it. What kind of energy machine can power particles to travel to such distances and radiate?

Any modern theory of radio sources must take account of their double structure, the central explosion, and the large energy reservoir needed for keeping the sources radiating. Before we consider possible scenarios, let us take note of another class of even more remarkable objects.

QUASI-STELLAR OBJECTS

In the early days of radioastronomy it became clear from the example of Cygnus A that considerable progress in the understanding of radio sources could be made by their optical identification. This process involves locating an object, with the help of optical telescopes, in a region close enough to the radio source that one can argue that the radio object and the optical object relate to the same system. For this process to succeed, the positions of both the objects must be known with good accuracy.

In the early 1960s, occultation of the radio source by the Moon was tried as a means of measuring the position of the source. The Moon's path is known with great accuracy and the process of occultation, by producing a clear-cut drop in the intensity of the source, helps in locating the position of the source behind the Moon. This was the method used in 1962 with the help of the radio telescope in Parkes, Australia, by Cyril Hazard. He and his colleagues M.B. Mackey and A.J. Shimmins succeeded in locating precisely the position of the radio source 3C 273 (273rd source in the 3rd Cambridge Catalogue). This was a key observation and, realizing its potential importance, the observers carried the data *in duplicate* on two separate flights from Parkes to Sydney, just in case . . .! The optical identification of 3C 273 then became possible and the optical object found in the vicinity of the source had a star-like appearance (see Figure 5.28).

In fact, the source had earlier been mistaken for a radio star in our Galaxy. Its extraordinary nature became apparent only when Maarten Schmidt at the Hale Observatories in California examined its spectrum. The spectrum was quite *unlike that for a normal star* in that it showed emission lines at a substantial redshift and so, on the basis of his analysis, Schmidt concluded that 3C 273 was located way beyond our Galaxy and was at least a *million times* as massive as a typical star such as the Sun. We defer until Chapter 7 our discussion on Hubble's law, with which Schmidt related the redshift to distance.

This object and another radio source, 3C 48, were the first of a new class of astronomical objects discovered in 1963. Both were star-like in appearance yet much more massive than stars, with spectral peculiarities which placed them very much farther away than stars in the Galaxy, and both were emitters of radio waves. These objects were

Figure 5.28: 3C 273, the first quasar to be identified, is unusually luminous. This montage illustrates the original picture on the Palomar Sky Survey plates, which showed a hint of a jet, a more recent image taken by ESO's New Technology Telescope in La Silla, Chile, clearly showing the jet and the fuzzy glow around the bright nucleus, and the Hubble Space Telescope images detailing the complex structure of the jet. (Courtesy of Herman-Joseph Roeser.)

called *quasi-stellar radio sources*, a term subsequently shortened to *quasars*. They are now also called *quasi-stellar objects* for the following season.

Although radioastronomy first led to the discovery of quasars, it soon became clear that not all quasars are radio sources. A number of radio-quiet objects, resembling in other respects the early quasars 3C 273 and 3C 248, were discovered and by now, with more than 7000 quasars known, it is estimated that the property of radio emission may be found in only a small percentage of all quasars.

However, studies of quasars have shown that they also tend to be strong emitters of X-rays. The general picture that emerges is that the X-ray emission comes from the most compact, innermost, region, the optical emission is from the intermediate region, while the radio emission (where present), is from an extended outer region. The implication therefore is that the main source of energy is at the centre, which is very compact. It could be a *supermassive black hole* formed by gravitational collapse.

This conclusion as such is not new. As far back as 1963, F. Hoyle and W.A. Fowler put forward the idea that the gravitational collapse of a supermassive object would give rise to a strong radio source. In the late 1960s, Philip Morrison and Alfonso Cavaliere had suggested the idea of a spinar: a supermassive rotating object (like a neutron star but a hundred million times more massive!) whose gravitational energy is converted to rotational and magnetic energy and thence to radiation.

That gravity *can* supply the requisite energy is known; the problem for the theoreticians has been to cook up a credible scenario whereby the gravitational energy is converted to electromagnetic radiation in an efficient manner!

In 1974, Martin Rees and Roger Blandford proposed another version of this process, which involves accretion onto a rotating supermassive black hole. Estimates from the observed data on radiation show that the black hole mass is around a *billion solar masses*. The efficiency of energy conversion from gravity to electromagnetic radiation is claimed to be as high as 20 per cent. (Compare this with the efficiency of energy conversion in the hydrogen-burning stage of a main sequence star, which is only 0.7 per cent). It is doubtful whether such high efficiency is really possible.

The problem remains of how we get the exact alignment and the double structure of Figure 5.27. In the mid-1970s, Rees and Blandford proposed a 'twin exhaust' model, illustrated in Figure 5.29. In this model plasma is squirted out from the flat disc in the directions of least resistance. For a rotating system these directions lie along the axis of rotation. The flow pattern of the two jets in Figure 5.29 is like that of jet engines, known in aerodynamics as the de Laval nozzle. The plasma is squirted out in these two opposite directions and proceeds until it encounters resistance from the interstellar medium; this limits the distance it can go and hence the size of the double radio source. The radio-emitting lobes are

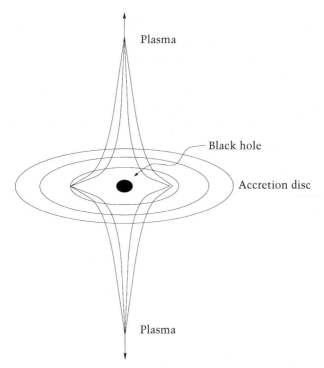

Figure 5.29: The twin exhaust model. Plasma is squirted out along two nozzles aligned with the axis of rotation.

explained as regions where a magnetic field is present, leading to radiation by the charged particles moving across it. Computer-image-processed pictures of radio sources by modern telescopes do show jets (see Figure 5.30).

A view held by many astrophysicists is that quasars are linked to the central nuclear regions of galaxies in an evolutionary process. As an example of an *active galaxy*, we see in Figure 5.31 the nucleus of M87, which is considerably brighter than its outer parts. The figure also shows a jet coming out. There are other types of galaxies, known as Seyfert galaxies (see for example, Figure 5.32), in which the contrast between the bright nucleus and its fainter surroundings is even more marked than it is in M87. Quasars may be one step further on this sequence so that if they are located at large distances we only see the bright central nucleus and nothing of the faint periphery (if it exists at all).

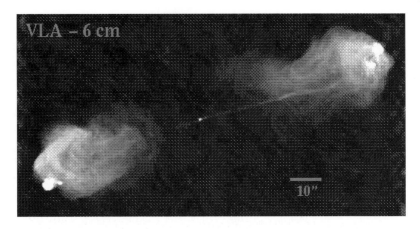

Figure 5.30: A radio map taken by the Very Large Array, New Mexico, of the radio source Cygnus A shows its double lobe structure and faint narrow jets from the centre to the lobes. (By courtesy of R. Perley, C. Carilli and J. Dreher, National Radio Astronomy Observatories.)

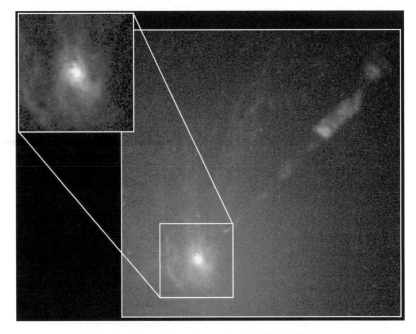

Figure 5.31: The galaxy M87 with active nucleus highlighted is shown on the left. The right panel shows the larger region containing a jet issuing from the nucleus (photograph taken with the Hubble Space Telescope Wide Field Planetary Camera 2, by Holland Ford *et al.*, NASA, STScI).

Figure 5.32: The Seyfert galaxy NGC 1068, which has an active bright nucleus (CCD image obtained by Joel Aycock using the guider on the Keck Telescope).

CONCLUSION

There we end our discussion of Wonder 5, which tells us the remarkable role played by the phenomenon of gravitation on the cosmic scale. On the scale of the atom, gravity pales into insignificance, so much so that atomic and particle physicists do not need to take heed of its existence when formulating their theories of the structure of the atom. Yet on the cosmic sale, gravity comes into its own as *the* controlling agent, whether it is on the scale of stars, radio sources or quasars. For, the accumulation of matter in a region beyond a critical scale triggers off a gravitational collapse that brings about highly condensed objects like black holes. Evidence showing outbursts of high energy in the cosmos tell us that gravity is making its presence felt. The challenge lies in figuring out the details.

Illusions in space (6)

IS SEEING, BELIEVING?

The science of astronomy has developed through observations of the cosmos. The motions of planets, the shining Sun and the twinkling stars, the dramatic supernovae, the clockwork-like pulsars, the powerful energy sources in quasars have all provided bread and butter to the astronomer and challenges to the astrophysicist. For, the latter has to explain what has been observed by the former. The law of gravitation, the phenomenon of thermonuclear fusion, the workings of the electromagnetic force at high energies, the behaviour of black holes, etc. are the products of the astrophysicists' responses to such challenges. And from these theories emerge new predictions of what the astronomer should look for in the cosmos.

While this unending cycle of observation → theory → observation → . . . continues, let us look at another aspect that has entered the field, an aspect that might possibly complicate what may have been a too-simple assumption in astronomy.

The assumption is: 'Seeing is believing'.

That is, astronomers must believe whatever their telescopes show; there could not possibly be anything 'wrong' with what they see on the photographic plates or the computer images.

Since the early 1970s this simple assumption has been called into question: *caution* is needed in interpreting astronomical images. In this chapter we will consider examples of 'illusions' warning the astronomer that there may be more to what they see than meets the eye.

Indeed, we have already referred to an example in Chapter 3. When we look at a star or a galaxy in an astronomical photograph we do not see it as it is *now*. Rather, we see it *as it was* when the light entering our camera today left it.

Take a look at the photograph of the great galaxy in Andromeda shown in Figure 6.1. A photograph is supposed to tell us what the object looks like. *Does the photograph in Figure 6.1 do that?*

The Andromeda galaxy is located at approximately two million light years from us. So we are seeing the galaxy as it was two million years ago, not as it is today. But even this statement is not strictly true, for a galaxy like Andromeda is about 100 000 light years across. So its two ends are not at the same distance from us. Depending on how the galaxy is orientated with respect to our line of sight, some parts of it could easily be 50 000 light years farther from us than some other parts. So we are not seeing them at the same epoch; the nearer part is being seen as it was 50 000 years *later* than the farther part. So what we are looking at is a mixture of different parts of the galaxy seen at different epochs. (Recall the mother–daughter photograph of Chapter 3.)

Figure 6.1: The Andromeda Nebula. The two ends of the galaxy are at different distances and so are being seen at different times.

SUPERLUMINAL MOTION IN QUASARS

In Chapter 5 we saw how the ideas of special relativity imply a fundamental limit on the speed of any material body. No such body can attain the speed of light, let alone exceed it. However, detailed observations of the internal structures of quasars began to reveal, in some cases, the existence of faster-than-light (superluminal) motion. We will take up this example as our next case of cosmic illusion.

Very-long-baseline interferometry

By the 1960s, radio telescopes were being developed in many countries on different continents. These were individual instruments designated for separate programmes. However, the radioastronomy community felt that much more could be gained by pooling their efforts. Very-long-baseline (VLBI) interferometry was one programme that emerged from this joint operation.

An interferometer is an instrument which makes use of the phenomenon of the *interference* of waves. A typical wave has crests and troughs. If a telescope is receiving two sets of waves from the same source, one of which has travelled a slightly longer distance than the other, owing to a different route, then their crests and troughs will not match. Depending on this difference in distances, or *path difference*, the crests of one may fall on the troughs of the other; the resulting wave, shown by the broken line I + II in Figure 6.2(a), is very small owing to this cancellation of the crests and troughs. If the path difference were increased by half a wavelength, the crests of one would then fall on the crests of the other. Correspondingly as shown in Figure 6.2(b), the resultant (nett displacement) of two nearly equal waves will change from nearly zero (crest on trough) to roughly double (crest on crest) the separate displacements. The technique of interference is therefore very useful in probing the structural details of the source since it reveals the differences in paths followed by the waves.

Figure 6.3 shows how the long baseline of a couple of telescopes linked together can provide a clearer picture of the source. In the figure we have two telescopes A, B receiving waves from the same source S. The spherical crests and troughs are shown as alternate solid and broken circles emanating from S. If A and B are each able to record an arrival time for the same wavefront, they can locate the source S more precisely. The longer the 'baseline' AB the greater the accuracy.

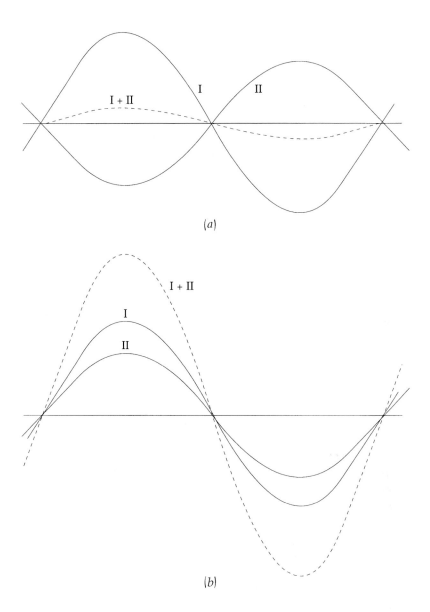

(a)

(b)

Figure 6.2: In (a) we see the destructive interference of two waves, whose resulting displacement, shown by the broken line, is small. If the waves were exactly equal and opposite the resultant would be zero. In (b) the crests of the two waves coincide and the resultant is large, roughly double the separate displacements.

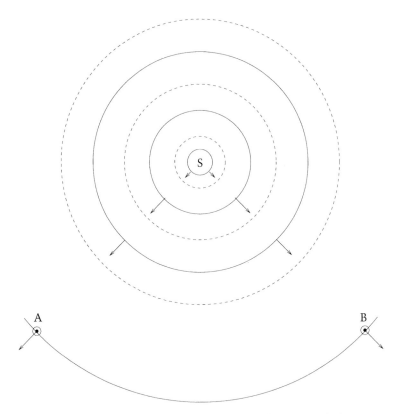

Figure 6.3: From source S spherical waves emerge, travelling outwards along expanding spheres. Alternate spheres with continuous and broken lines indicate respectively the crests and troughs of the waves. Two distant telescopes at A and B may detect the same wavefront.

However, normally one expects A and B to be linked by cables, so that their observations can be collated. If A and B are too far apart such connections are not possible. Nevertheless, because of the availability of extremely accurate atomic clocks, the cable-linking can be dispensed with. Observers at A and B can keep precise timings of the crests and troughs passing them and thereby achieve the same end. This feature has made the concept of a very-long-baseline interferometer realizable even with A and B separated by thousands of kilometres. This in turn makes it possible to achieve a very high resolution of a quasar source. That is, even if its two components are separated in direction to a terrestrial observer by an angle less than a *thousandth part of a second of arc*, they can be seen distinctly.

Just how good this resolution is can be seen by likening it to seeing two ends of a pen as separate points from a distance of 2000 kilometres!

Given the high resolution achievable by the VLBI technique, scientists have applied it to those quasars which are strong emitters of radio waves. (Of the total population of quasars only around ten per cent emit strongly at radio wavelengths.) The technique has revealed details of the structures in quasars on the scale of a few light years.

Let us pause here to see how different techniques have provided the astronomer with increasing resolution.

Figure 6.4 shows a series of maps of a radio source associated with the galaxy NGC 6251. The figure at the top shows the size of the source on the very largest scale of *a million light years*. The higher-resolution map in the middle reveals the existence of a jet about 500 000 light years long but with structure noticeable on the scale of 10 000 light years. These details are further enhanced by VLBI at the bottom with structure seen on the scale of one light year. (The entire structure at the bottom is less than 10 light years long!)

This series of maps may be compared with the map of a country, then of a city and finally that of a house. Each map shows a greater resolution than the preceding one.

The motion of the VLBI components of a quasar

In the early 1970s, astronomers were able to study the high resolution maps of quasars taken at different times. Figure 6.5 shows a map of the quasar 3C 345 taken in 1974. Such maps have been taken at almost yearly intervals and appear similar in structure.

The maps are typically contour maps which have lines of constant intensity. Comparing with the contour maps in a geography atlas, where mountains of different heights can be identified from their contours, we see that the map in Figure 6.5 shows two 'peaks', say A and B, of high intensity in each map. Also, if we look at the same two peaks in all such maps in chronological order, the implication is that A and B are moving apart. How fast is this relative motion?

Let us now see how this speed is measured. For this we need to know the length AB in each of the observations. What the astronomer can measure directly is the angle subtended by the arc AB at an observer O. Figure 6.6 illustrates the problem.

Here we have a circle with O as centre on which both A and B lie. The angle AOB can be measured: it is the angle between the directions of A

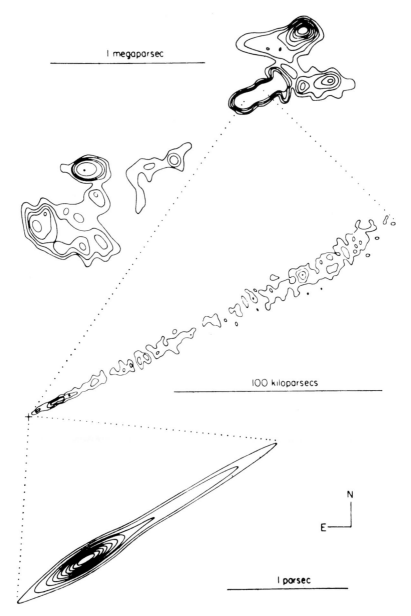

Figure 6.4: A hierarchy of resolution reveals structure in the radio source associated with NGC 6251: at different sizes. One parsec is approximately three light years. (Figure based on the work of A.C.S. Readhead, M.H. Cohen and R.D. Blandford, *Nature*, **272**, 131, 1978.)

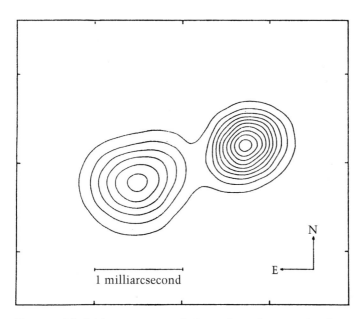

Figure 6.5: The brightness contours of 3C 345 observed at a wavelength 2.8 cm, midway through the year 1974. (From the work of D.B. Shaffer *et al.*, 1977.) The intensity peaks are called *A* and *B* in the text.

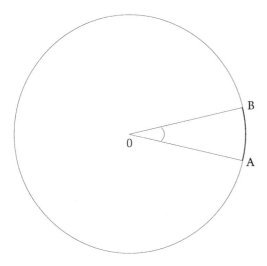

Figure 6.6: The relationship between the arc length *AB* and the angle *AOB* depends on the radius of the circle.

and B as seen from O. We wish to measure the *length* of arc AB, given the radius of the circle. Let us suppose the radius is R.

Now we know that the circumference of the circle is $2\pi R$. The total angle described by the entire perimeter of the circle around O is $360°$. And if we take only a fraction of the perimeter, such as the arc AB, then the angle subtended at O by this arc will bear the same ratio to $360°$ as the length AB bears to $2\pi R$. So, using simple proportion, AB is $2\pi R \times$ angle $AOB/360°$.

So, provided we know the distance R of the quasar, we can estimate the length AB. The distance R is provided by the redshift z of the quasar and Hubble's law of the expansion of the universe. We refer the reader to Chapter 7 for these details. For the time being we simply assume that the redshift of the quasar can be measured from its spectrum and that the distance of the quasar from us is obtained by multiplying the redshift by a fixed number, which is of the order of 10 billion light years. Thus, in the case of 3C 345, its redshift being 0.595, the distance is about six billion light years.

Applying this method, astronomers could measure AB for each of their successive maps of 3C 345. The distance was found to increase year by year. In Figure 6.7 we see a plot of the separation AB against time measured in years.

This is where a most surprising result was first noticed. *The length AB was found to increase at a rate of about 3–8 times the speed of light.* This *superluminal motion* is clearly against the dictates of special relativity.

The case of 3C 345 was not an isolated one. VLBI studies of a number of other quasars revealed similar results. And thus the astronomers could not dismiss them as flukes or experimental errors. They had to be explained.

THREE EXPLANATIONS OF SUPERLUMINAL MOTION

Of course, one easy way out at once suggests itself: we get a large value of the separation speed for AB because the length AB is too large. The length AB is too large because the distance R of the quasar is too large. If, for example, R were shorter by a factor 100, the separation speed of A and B would drop by the same factor and this would remove any conflict with the special theory of relativity.

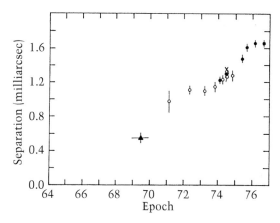

Figure 6.7: A chronological plot of how the angle AOB has increased for 3C 345 between 1969 and 1977. Measurements were made at the following wavelengths: ×, 2.0 cm; •, 2.8 cm; ○, 3.8 cm; ▲, 6.0 cm. (Based on the work of K.I. Kellermann and D.B. Shaffer, reported in *Proceedings of the Colloquium on Evolution of the Galaxies and Its Cosmological Implications*, C. Balkowski and B.E. Westerlund eds., CNRS, Paris, 1977.)

This soft option, however, is not popular with the astronomical community. For, it would mean that the method of deducing the distance of the quasar from its redshift would have to be wrong. We will refer to the controversy surrounding quasar distances in the Epilogue. Here we will simply take the view, shared by the majority of the astronomical community, that quasar distances are indeed related to their redshifts according to Hubble's law. In short, our method of estimating R is correct. So that eliminates the easy way out.

The Christmas tree model

Think of a Christmas tree with tiny electric lights draped around it. The lights can have a sequence of 'on–off' switching built into their wiring so that when the power is turned on they follow that sequence. For someone looking from a distance the illusion of motion is created. Neon lights also create an illusion of motion in large advertising panels on street corners.

Based on these examples, the idea emerged that in these quasars too we are not seeing a physical motion of components but are instead witnessing *different* components being 'lit up', as shown, for example in the lighting sequence of Figure 6.8.

If we shine a narrow beam from an electric torch onto the wall of a dark room, we can generate motion of the light patch across the wall by moving the beam. Such motions can be made arbitrarily fast and they do not conflict with special relativity as they do not describe motions of material bodies.

However, such attempts to understand the apparent superluminal motions in quasars with complex geometrical structure began to look more and more contrived as data for more of these sources began to accumulate.

The beaming model

In this model, based on an earlier idea of Martin Rees from Cambridge, UK, the illusion of superluminal motion is created as follows.

Imagine (see Figure 6.9) a source consisting of two radiating sources *A* and *B*. Source *A* is fixed relative to the observer *O* whilst *B* is moving towards *O* in a direction such that the line *AB* is almost pointing at *O*. Hence the name 'beaming' for this model. In the figure as drawn it would appear that *O* is seeing both *A* and *B* at the same epoch (moment of time).

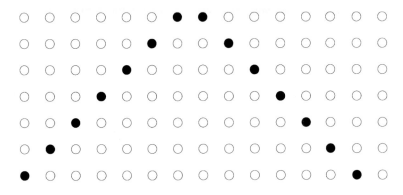

Figure 6.8: We see a time sequence in which blobs farther out from the centre get lit up progressively. Although all blobs are stationary, the illusion is created that there are only two illuminating blobs moving outward. The time sequence is downwards.

Figure 6.9: The scenario of the beaming model. See text for how the model works.

This is incorrect when we notice that their distances from O are different: A is farther from O than B. So the observer sees A at *an earlier epoch* than B. Therefore the length AB as estimated by O at any epoch is *longer* than the actual length. Under these circumstances the apparent speed of separation of B from A will exceed the real speed.

If we take this method of resolving the problem as correct we are essentially arguing along the lines used for interpreting the photograph of the Andromeda galaxy. Recall that the photograph is really a composite of different parts of the galaxy at different epochs. In the same way, Martin Rees's explanation is based on the possibility that we are observing A and B at different epochs. Assuming that A is fixed and B is moving towards us, light from B to us has to cover an increasingly shorter distance with time. We have to allow for this progressively growing time delay between the signals from A and B. That is why our estimate of their separation speed turns out to be wrong: we get an inflated answer which could exceed the speed of light in certain special situations.

The 'certain special situations' require the line AB to be almost aligned with respect to the observer O. That is, the angle AOB has to be very small, of the order of a few degrees. That is why, it is argued, superluminal motion is seen in only a handful of the large population of radio quasars.

I will return to the 'illusion' of superluminality towards the end of the chapter, with another possible explanation.

For the time being let us turn to another 'illusion'.

THE BENDING OF LIGHT

While Isaac Newton was working on issues related to light, he had conjectured whether light is attracted to matter by the latter's gravitational force. He asked

> 'Do not Bodies act upon Light at a distance, and by their action bend its Rays; and is not this action (caeteris paribus) strongest at the least distance?'
>
> *Opticks, Query 1*

That the possibility that light is bent by gravity occurred to Newton is not surprising, considering his intuitive genius and also his belief that light is made of particles (which he called *corpuscles*). However, he had no experimental or observational means to settle his conjecture and so he left the matter there.

A 'Newtonian' calculation

We can, however, use Newtonian ideas to work out how much bending, if any, light would suffer if it passed close to a massive body. Figure 6.10

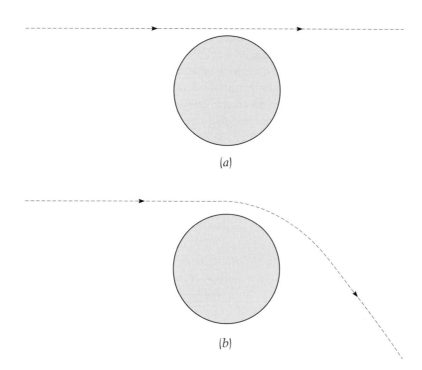

Figure 6.10: The two possible directions of light rays considered as swarms of particles, emerging from the vicinity of a massive sphere are shown by (a) and (b). In (a) light is unaffected by gravitation while in (b) it is so affected.

illustrates the situation; the assumption is that the photons, the packets of light, are attracted to the massive body according to the Newtonian law of gravitation. Thus we attribute a mass m to the photon by setting its energy equal to mc^2.

The figure shows a swarm of particles coming from a great distance, with the speed of light. As they approach the massive sphere, two possibilities arise. (a) The particles are immune to the force of gravity of the sphere and so continue along the straight line. (b) They *are* attracted by it, bend round it and emerge in another direction.

We may consider (a) and (b) as the possible outcomes of Newton's conjecture about light. In the second alternative the bending can be calculated as the angle between the two emerging directions in (a) and (b). The answer is $2GM/c^2$, where G is the Newtonian gravitational constant, M the mass of the sphere and c the speed of light. This gives the bending angle in radians.[10]

If we work this out for a light ray grazing the surface of the Sun the answer is the very small angle of about 0.87 seconds of arc. This value is indicative of the rather weak effect Newtonian gravity has on light, at least in the environment of our solar system, wherein the Sun is by far the strongest gravitating body.

The bending of light in general relativity

We now return to the point we deferred in the previous chapter. How is the path of light affected by gravity according to Einstein's general theory of relativity?

Recall that in relativity theory, gravitation is not looked upon as a force in the Newtonian sense. Rather, it is identified through its effect on the spacetime geometry. So to solve the problem analogous to the Newtonian one which we just looked at, we first find out the non-Euclidean geometry in the vicinity of the gravitating sphere and then work out the path of a light ray in such a spacetime.

The first path of the problem we have already discussed in Chapter 5. Karl Schwarzschild showed how to determine the spacetime geometry outside a massive sphere. The second part is then straightforward. We have to determine the *null geodesics* in such a spacetime.

We encountered geodesics as the curved spacetime equivalents of straight lines connecting two spacetime points (see Chapter 5). For material particles these are the worldlines describing 'uniform motion in a straight line'. Material particles follow these trajectories when under the gravitational effects that have made the spacetime non-Euclidean. So light also has to follow such worldlines, with the added proviso that they are null lines.[11]

Unlike the Newtonian case, there is no ambiguity here. Travelling in a spacetime with geometry determined by gravity, *light has to modify its trajectory*. Judged by the Euclidean criterion for 'straightness', light

---

[10]To convert radians into seconds of arc a rough and ready rule is to multiply by 200 000. In Figure 6.6 the ratio AB/OA gives the angle in radians. If $AB = OA$ then the angle equals one radian.

[11]We remind the reader that along a null line the separation between any two points is zero when measured according to the rules of relativity (see the subsection entitled 'The speed of light' in Chapter 5).

rays bend. A more correct statement is that given the prevalent non-Euclidean geometry, the 'straight-line' path followed by light according to the rules of this geometry will be different from the Euclidean path.

However, since we are comparing the relativistic result with Newton's alternative (b), see the previous subsection, we will use the rather loose but more commonly used phrase 'the bending of light'.

How much does light bend? Having seen the Newtonian answer, the relativistic one is easy to state: *it is exactly double the Newtonian value.* In other words, the bending of light passing close to the Sun will be by an angle 1.74 seconds of arc.

Although the exact details of the spacetime geometry around a spherical mass became available in 1916 after the work of Schwarzschild, Einstein himself had worked out the relativistic bending of light in 1915, soon after formulating his equations of gravitation. In those early days very few scientists really understood what general relativity was all about. Most had found the notion of a non-Euclidean geometry that actually applied to spacetime to be very bizarre and counter-intuitive.

A.S. Eddington, however, was one of those few who did grasp the essence of general relativity. Being an astronomer who had gone through the rigours of the Cambridge Mathematical Tripos,[12] Eddington could not only appreciate the mathematical elegance of general relativity but, with his astronomical background, also think of an astronomical test of the bending of light.

The 1919 eclipse expedition

Figure 6.11 is an adaptation of the second possiblity of Figure 6.10. It shows a star A whose light grazes the surface of the Sun before coming towards the observer. The observer therefore sees the star image at A' instead of at A; that is, the star's image jumps from its regular position *if it happens to be just behind the Sun.*

The expected shift in the direction of the star is no more than about 1.7 seconds of arc. However, there is a practical difficulty: how can we see the star with a dazzlingly bright Sun in the forefront? This is possible only at the time of a total solar eclipse.

[12]Eddington was a Senior Wrangler of the 1904 vintage.

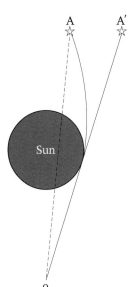

Figure 6.11: The dotted line is the Euclidean straight line along which light from the star A travels to the observer O, when the Sun is not anywhere near this line of sight. If, however, the Sun comes in the way, its gravitation modifies the geometry in its neighbourhood, making it non-Euclidean. The light from A then 'bends' as shown by the continuous line, and the observer O sees the image of A at A' along the tangent to the light path at O.

Realizing this, Eddington proposed a measurement of this phenomenon at the time of the total solar eclipse of 29 May 1919. A grant of £1000 obtained by the Astronomer Royal, Sir Frank Dyson, made this project possible. Two teams set out: one, consisting of Eddington himself and E.T. Cottingham, went to the island of Principe in the Gulf of Guinea; the other, consisting of C.R. Davidson and A.C.D. Crommelin, went to Sobral in Brazil.

In the end (for solar eclipse observations can be a chancy affair!), both teams were rewarded with perfect seeing conditions and measurements could be carried out.

The results of the observations were announced by Sir Frank Dyson on 6 November 1919, at a crowded joint meeting of the Royal Society and the Royal Astronomical Society. There had been great excitement and expectation about what the findings would be. Would light show any bending at all? Would it bend as calculated by the Newtonian methods? Or, would the answer favour relativity? A.N. Whitehead, who attended the meeting, captured the scene graphically in these words:

The whole atmosphere of intense interest was exactly that of a Greek drama: we were the chorus commenting on the decree of destiny as disclosed in the development of a supreme incident. There was . . . in the background the picture of Newton to remind us that the greatest of scientific generalizations was now, after more than two centuries, to receive its first modification . . .

The results did favour general relativity. Within the estimated errors, the bending of light was closer to the value 1.74 seconds of arc than to half this value, as obtained from Newtonian gravity.

The success of the eclipse expedition made Einstein an instant celebrity. Although the concept of curved spacetime still remained beyond the comprehension of most people, the results confirmed that nature did seem to follow these apparently crazy ideas.

And, of course, it was the first indication to the astronomer that because of the bending of light by 'en-route' masses, the observed positions of images in the sky may not quite represent reality.

But several decades had to pass for this to sink in.

A diversion

Eddington himself had been greatly thrilled by the total solar eclipse and was inspired to write a parody of the famous Rubaiyat:

Ah Moon of my Delight far on the wane,
The Moon of Heaven has reached the Node again
But clouds are massing in the gloomy sky
O'er this same Island, where we laboured long – in vain?
And this I know; whether EINSTEIN is right
Or all his Theories are exploded quite,
One glimpse of stars amid the Darkness caught
Better than hours of toil by Candle-light
Ah Friend! could thou and I with LLOYDS insure
For God this sorry Coelostat so poor,
Would we not shatter it to bits – and for
The next Eclipse a trustier Clock procure
The Clock no question makes of Fasts or Slows,
But steadily and with a constant Rate it goes,
And Lo! the clouds are parting and the Sun
A crescent glimmering on the screen – It shows! It shows!!
Five Minutes, not a moment left to waste,
Five Minutes, for the picture to be traced –
The Stars are shining, and coronal light
Streams from the Orb of Darkness – Oh make haste!

For in and out, above, about, below
'Tis nothing but a magic Shadow show
Played in a Box, whose Candle is the Sun
Round which we phantom figures come and go
Oh leave the Wise our measures to collate
One thing at least is certain, LIGHT has WEIGHT
One thing is certain, and the rest debate –
Light-rays, when near the Sun, DO NOT GO STRAIGHT.

But there was another side to the expedition.

The 1919 solar eclipse expedition that provided the first observational evidence for the bending of light by gravity also had a byproduct: Eddington's problem in probability theory.

As explained, above, four observers were deputed to make the eclipse measurements: Davidson and Crommelin went to observe in Sobral in Brazil while Cottingham and Eddington went to the island of Principe in the Gulf of Guinea. A great deal had depended on their results. Would light show any bending by gravity? Would the effect be as predicted by the (hybrid) Newtonian gravity or double that value, as predicted by Einstein?

In an after-dinner speech before they set out, Crommelin referred to the four observers C, C', D and E and hinted that a problem might arise as to the veracity of their claims with so much at stake and given the possibility that each of them might be tempted to bend the truth once in a while! The question was subsequently restated by Eddington in the following form.

A, B, C and D each speak the truth once in three times (independently). D makes a statement and A affirms that B denies that C declares that D is lying. What is the probability that D was speaking the truth?

What answer do you get?

This convoluted problem is solved by the use of arguments based on conditional probability. Try to solve it, if you like! The answer is that the probability that D was telling the truth is 25/71.

A postscript

In retrospect, however, astronomers now concede that the results of the 1919 expedition were not really as definitive as claimed at that time, because the experimental errors had been underestimated.

Indeed, any experimental result based on several measurements is inevitably subject to experimental inaccuracies over which the experimentalist has no control. These inaccuracies can creep in for several reasons. The measuring apparatus has a limited sensitivity. For example, a metre rod with divisions of one millimetre cannot make measurements which have greater accuracy than one millimetre. The optical equipment used in the 1919 expedition had limited accuracy.

Then there may be random errors which cause individual measurements to exceed or be less than the average. Such errors arise when comparing the photographs of the star field with and without the Sun. But then there are also systematic errors caused by additional effects which were not realized and accounted for at the time of the experiment.

In the case of the 1919 eclipse expedition, one systematic effect not accounted for was the bending of light caused by refraction, when light passes through a variable medium. (We encountered this effect in Chapter 1.) The Sun has an atmosphere surrounding it and a light ray passing obliquely through the atmosphere will get bent because of the variations in the density and temperature of the medium it passes through. Figure 6.12 shows this effect in an exaggerated form. For optical wavelengths the effect is small, but any claim about the bending of light by the gravity of the Sun can be made only after making allowances for the above effect.

In similar optical experiments during later eclipses also the accuracy remained limited. Although one could demonstrate that the bending of light was closer to the relativistic value than to the Newtonian one, this could not be confirmed with high accuracy. It was only in the 1970s that a solution was found to the problem: *use microwaves instead of visible light*. This has three advantages.

First, the bending due to refraction of microwaves can be estimated and accounted for by making simultaneous measurements at two wavelengths, and therefore it does not confuse the issue. Second, the Sun itself is not bright in microwaves. So if we have a strong source of microwaves in the background, we can carry out the experiment *without having to wait for a total solar eclipse*.

Realizing these implications, radioastronomers using waves in the wavelength range 10–40 cm looked at the change in the direction of the quasar 3C 279 when the Sun happened to pass across its line of sight. The shift in position of the quasar could be measured in relation to another nearby quasar, 3C 373.

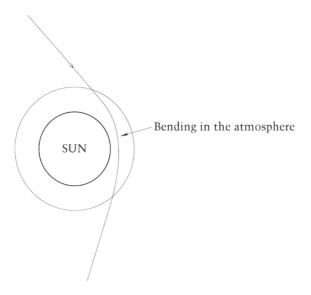

Figure 6.12: The bending due to refraction of light during its passage through the solar atmosphere, shown in an exaggerated form.

Finally, radiointerferometric measurement techniques were much more accurate than those available to optical astronomers, and these experiments had therefore very small errors. The conclusion was unmistakably in favour of general relativity, within an experimental error of one per cent.

GRAVITATIONAL LENSING

Figure 6.13 shows an ordinary lens, the type used in a magnifying glass. The ray diagram shows how the magnification occurs. The rays emerging from the object AB appear to come from a much larger source $A'B'$, which is the 'virtual' image of AB.

Lenses are of various kinds. The one in Figure 6.13 has both its surfaces *convex*. There are others with both surfaces *concave* or with one surface convex and the other concave. All of them form images of real objects by bending light rays suitably. The cause of bending here is, of course, *refraction*.

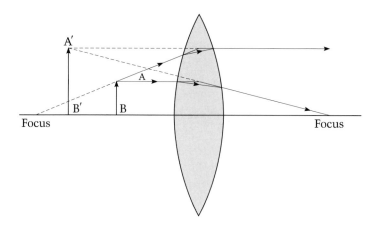

Figure 6.13: A convex lens. The rays of light are drawn to illustrate how a virtual image of a source AB is formed at $A'B'$.

Since gravitation can also bend rays of light, can we likewise encounter situations where the phenomenon produces lenses? This issue was first raised in the 1930s by the California Institute of Technology astronomer Fritz Zwicky. He wrote in 1937:

> Last summer Dr V.K. Zworykin (to whom the same idea had been suggested by Mr Mandl) mentioned to me the possibility of an image formation through the action of gravitational fields. As a consequence I made some calculations which show that extragalactic nebulae offer a much better chance than stars for the observation of gravitational lens effects.

Zwicky had proposed the use of lensing to detect dark matter, i.e., matter that cannot be seen but whose gravitational influence can nevertheless bend the light from visible matter. We will refer to dark matter again in the next chapter.

It happens on some rare occasions that a highly perceptive idea remains neglected because the scientific community is not ready for it. Zwicky's ideas were three to four decades ahead of their time and were recalled and appreciated much later, when he himself was no more.

More or less the same fate awaited the work of S. Refsdal and Jeno Barnothy in the mid-1960s. Independently, these scientists explored the possibility of lensing by galactic masses and its effect on images of quasars, which were then beginning to be discovered (see Chapter 5). Their ideas were considered interesting curiosities far from the fashionable lines of research.

Then in 1978–9 the subject erupted all of a sudden, catapulting gravitational lensing onto centre stage. Zwicky's prophecy expressed in 1937 became realized:

> Provided that our present estimates of the masses of cluster nebulae are correct, the probability that nebulae which act as gravitational lenses will be found becomes practically a certainty.

DISCOVERY OF THE FIRST GRAVITATIONAL LENS

The announcement of what could be the first example of a gravitational lens was made in *Nature* by three astronomers, Dennis Walsh from the Nuffield Radio Astronomy Laboratories at Jodrell Bank, UK, together with Bob Carswell from the Institute of Astronomy, Cambridge and Ray Weymann from the Steward Observatory of the University of Arizona. The announcement generated considerable excitement and discussion. Being the first case of its kind, astronomers were naturally cautious in accepting the interpretation proposed by these authors.

Before we go into the details of this supposed 'lens' and its theoretical interpretation, let us first take a brief look at its history. For, as narrated by Walsh, the path to its discovery was not a straight and narrow one but rather a meandering one with several chance turnings that need not have taken place!

The story, as told by Walsh a few years later at a conference on gravitational lensing, began in the early 1970s when the Mark I telescope at Jodrell Bank (see Figure 6.14) was upgraded. Bernard Lovell, the director, asked workers at the laboratories to submit new proposals for observing with the new dish.

At the time, the science of radioastronomy was entering a new phase where improvements in technology would enable astronomers to make more detailed studies of radio sources and also to pinpoint the positions of radio sources more accurately in the sky. As we saw in the previous chapter, the accurate position of the source 3C 273 had been determined by lunar occultation and this had enabled optical astronomers to 'see' the source, that is, to identify it optically. In this way a new class

Figure 6.14: The 76-m dish of the Lovell Telescope at Jodrell Bank. (Photograph by courtesy of Prof. R. Davies, Director of Nuffield Radio Astronomy Observatories, Jodrell Bank.)

of radio sources, called quasi-stellar radio sources, had been discovered. So, much was to be gained by obtaining very accurate positions of the many radio sources that had still not been identified.

The process of optical identification involves looking for an optical source within the error rectangle of the radio position. Normally, there may be several sources in the rectangle and further diagnostics such as the spectrum of the source may be required for the optical astronomer to be certain that this is indeed the source observed in the radio. The more accurate the radio position, the smaller is the error rectangle and the job of identification becomes easier and more definitive.

Dennis Walsh suggested the use of the upgraded telescope, together with the Mark II dish of 25 metres, to make up an interferometer with improved resolution. (We have discussed this technique in connection with the VLBI earlier in this chapter.) This would allow more accurate determinations of the positions of radio sources in the sky, which in turn would help in their optical identification.

Walsh, together with Ted Daintree, Ian Browne and Richard Porcas, began observing in November 1972 with an initial time allotment of one month. However, several difficulties intervened and they could not complete the job in time. Bernard Lovell, who had been greatly enthused by the results obtained up to then, came to their help and in his capacity as the director of the Observatory, granted them an additional month for continuing with the operation. This was the first of a series of fortuitous circumstances that helped find the lens.

On 4 January 1973 they detected a radio source to which they gave the catalogue number 0958+56. Figure 6.15 shows the bump in intensity in the scans which suggested to these observers that here was a new radio source. This source was destined to play a key role in the discovery of the first gravitational lens.

The next step towards optical identification of the source was to measure the position more accurately with the 300-foot dish at the National Radio Astronomy Observatory (NRAO) at Green Bank, USA; this was carried out by Richard Porcas, and Figure 6.16 shows what he found. In Figure 6.16(a) we see the Palomar Sky Survey print with the object, now numbered as 0957+561 and marked by two lines at right angles. The Sky Survey prints are extremely useful for identification, as they give a very extensive coverage of the sky, locating objects brighter than a specified limit. The more eye-catching object in this figure is the galaxy with catalogue number NGC 3079.

In Figure 6.16(b) is a radio map made in 1986 by Condon and Broderick with the NRAO's 300-foot dish, which shows the radio contours in that region. We see that in fact the source 0958+56 is the weaker companion of a stronger source near the galaxy NGC 3079. This stronger radio source had been known earlier and catalogued as

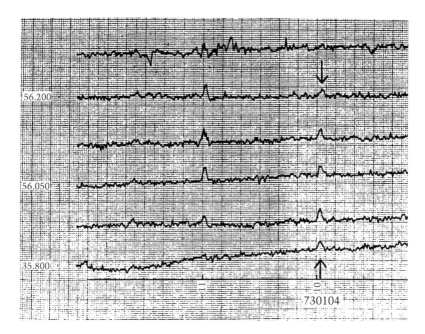

Figure 6.15: The first detection of the radio source 0958+56 was through the above survey charts. (From the article by D. Walsh, in *Gravitational Lenses*, J.M. Moran, J.N. Hewitt and K.Y. Lo eds., Lecture Notes in Physics **330**, Springer.)

4C 55.19. However, this is the story of 1986. What was the situation a decade earlier?

In 1976, Porcas had been aware of the 4C radio source. To locate the Jodrell source 0958+56 he proceeded to look north and eventually found it. Had he gone southwards he would not have found it, nor of course would he have done so had he stopped at detecting the stronger 4C source. This was another fortuitous circumstance.

Walsh had also commented on the relative nearness of the two radio sources and the rather unusual circumstance that they should in fact have detected the stronger source in the Jodrell survey (but did not) and not the weaker source (which they did)!

Next was the actual identification programme. Ann Cohen from Jodrell Bank and Meg Urry at NRAO were independently working on it and, by 1977, both had arrived at a likely blue star-like object as a

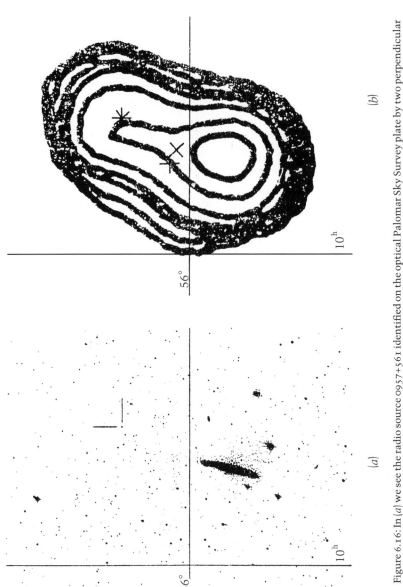

(a)

(b)

Figure 6.16: In (a) we see the radio source 0957+561 identified on the optical Palomar Sky Survey plate by two perpendicular lines. Notice the prominent galaxy NGC 3079 nearby. In (b) we see part of the (radio) atlas of Condon and Broderick, on the same scale as in (a). The radio contours peak around the galaxy NGC 3079. The position of 0957+561 is marked by (*). The Jodrell Bank position of the radio source, 0958+56, is shown by (+). Nearby is the stronger radio source 4C55.19, noted earlier by Richard Porcas, and marked (×). Notice that the Jodrell Bank position of the new radio source is about midway between the true position (*) and the NGC galaxy. (Attribution: same as for the previous figure.)

possible identification. Moreover, the object appeared to be a double source. However, its angular separation from the radio position given by Porcas was some 17 seconds of arc, which did not promise a likely identification. Nevertheless, a blue star-like object could still be a quasar, and Walsh and Carswell decided to observe it more carefully at the 2.1-metre telescope of the Kitt Peak National Observatory. The double source was catalogued as 0957+561 and when they took the spectra of the sources (see Figure 6.17) *they found them to be extremely similar;* so much so that they thought that they had made the mistake of taking the spectrum of the same object twice!

But careful re-examination showed that they had not erred; they really were looking at a close pair of quasars with identical spectra and identical redshifts of 1.4. The two images were separated by a small angle, six seconds of arc. This was in March 1979. Additional observations were needed to confirm this remarkable finding. And, as another lucky break, they met the astronomer Ray Weymann, who had come up to Kitt Peak for another observation on the Steward 2.3-metre telescope.

And what had brought Weymann there? He had been informed at short notice that he could have a single night on this telescope. Since nights at major telescopes are a precious commodity for observers, he had readily come to use this extra night. It so happened that he had been carrying out studies of quasars at redshifts in a range that included the redshift 1.4. So he agreed to devote his night to the new object. The observing conditions were superb, and Weymann's observations confirmed the earlier findings of Walsh and Carswell. The similarity between the two objects was so uncanny that, for the first time, the three of them dared think of the likelihood that they had observed a *gravitational lens.*

In a lighter vein, Walsh recalls a bet he took with another astronomer Derek Wills, who was an expert on close pairs of quasars. This was before the final work on 0957+561 was carried out. When Walsh asked Wills what he thought the blue stellar objects would turn out to be, he replied 'stars'; for this was the more likely possibility. So Walsh had a bet: he would pay Wills 25 cents if the pair turned out to be stellar, whereas if they turned out to be a pair of quasars, Wills would pay Walsh one dollar. At that stage Walsh thought it too facetious to say that Wills should pay a hundred dollars in the more unlikely event that the quasars turned out to have the same redshift!

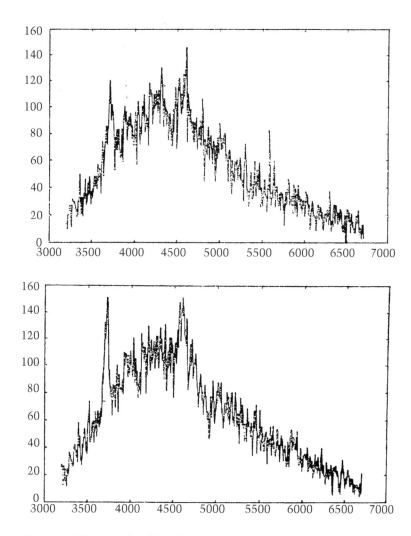

Figure 6.17: The spectral profiles of the two sources *A* and *B* in 0957+561 are very similar.

After the spectra confirmed that the blue objects were indeed quasars, Wills paid up the bet with a silver dollar. So when his sons asked in a sceptical vein what good was a gravitational lens to him, Walsh could reply: 'Well, I made money out of it.'

Details of the images

The discovery of the gravitationally lensed quasar 0957+561 described above was followed up by several investigations into different aspects of its lensing system. We will now present a few highlights.

Figure 6.18 shows on the left the optical images A, B of the 'double' quasar. They showed visual similarity, except for a small bump in the left-hand image, B. Thanks to the techniques of computerized image processing, the 'extra' part in B can be eliminated to make the two images identical. But what is that extra bump due to?

Further work showed that the extra bump is a galaxy of redshift 0.36. As explained in Chapter 7, the redshift is a measure of the distance of the galaxy from us. Based on Hubble's law, described there, we may take the distance of an extragalactic object as given by its redshift multiplied by a distance of about 10 billion light years.[13] Thus the redshift value 1.4 of the quasar images A, B indicated that the quasar is located at a distance of about 14 billion light years whereas the bump-galaxy is considerably closer, at 3.6 billion light years.

So this galaxy is situated en route to the quasar, although slightly off the direct line of sight, and this raised the exciting possibility that it

Figure 6.18: Hubble Space Telescope image of the quasar 0957+561A, B, which is believed to be an example of gravitational lensing (observations by Jean-Paul Kneib and Richard Ellis, courtesy of STScI).

[13]As we shall see in Chapter 7, the exact redshift–distance formula depends on the value of Hubble's constant and the cosmological model describing the universe.

may, in fact, be the lensing galaxy. Theoretical modelling of the image system lent support to this conjecture. Moreover, further details of the radio structure of the source required *additional* lensing by a cluster of galaxies housing the lensing galaxy.

The radio structure is shown in Figure 6.19. Again we see similarity in the images *A, B*. The radio lobes at these locations look identical. There are additional features, however, in the figure that do not match, image for image. To explain some of these additional features, the second (cluster) lens is required. Such a cluster would include the lensing galaxy itself.

Figure 6.20 shows a theoretical light ray diagram for 0957+561. Notice that the two images *A* and *B* are seen by the observer on the Earth through two light paths from the original source. Thus *neither A nor B is located at the real position of the source*; both are illusory. Nevertheless, mathematical modelling of the lens system can be used

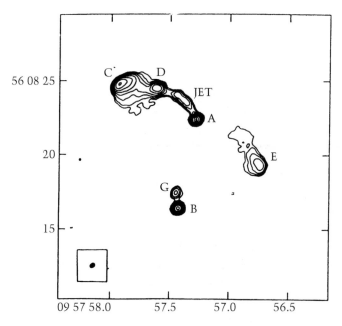

Figure 6.19: The radio map of 0957+561 taken with the Very Large Array, at a wavelength of 6 cm. It shows components *A* and *B* coinciding with the optical components; in addition there are other components *C, D* and *E* linked with the *A* component. Vertical axis, declination; horizontal axis, right ascension. (From D. Roberts *et al.*, *Astrophysical Journal*, **293**, 356, 1985.)

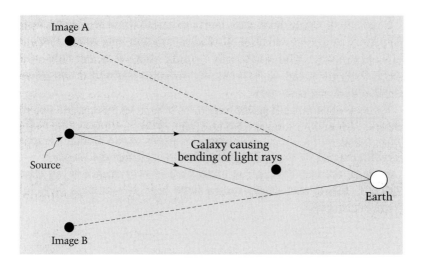

Figure 6.20: A ray diagram for gravitational lensing of the double quasar 0957+561A, B.

to estimate the relative brightness of the observed images. For example, image A is about a quarter brighter than image B, both in optical as well as radio, which means (see Figure 6.20) the rays forming image A take away and concentrate that much greater a part of the light from the original source than the rays forming image B. The mathematical model must also match the observed angular separation of the observed images, in this case six seconds of arc.

However, there is a clinching test that can tell us quite clearly whether such a gravitational lens is indeed responsible for what we see or whether the two images are straightforwardly two distinct sources which happen to have very similar spectra and shapes. This test is called the *time delay test* and it works as follows.

Suppose the source does not have constant brightness but has irregular ups and downs in its brightness profile, if monitored over a longish period. The same ups and downs will be noticed in A and B but not at the same time. For, the paths of light forming the two images are not of equal length. Since light will correspondingly take different time intervals to make the journey, we are not seeing A and B at the same time. So the ups and downs in the source will be seen in A and B at different times.

Hence, in the time delay test, we try to match these brightness fluctuations in images *A* and *B* by allowing a suitable time delay. Thus, if the model predicted that the path forming image *B* is one light year *longer* than for image *A*, then the fluctuation pattern of *A* should be repeated in *B one year later*.

Tests of such a time delay for 0957+561 have so far been rather inconclusive. Theoretical models expect a time delay of about a year and a quarter, depending on the geometrical features of the model. Clearly, the source needs to be monitored further to convince the sceptics.

In short, the very first pair of images found as examples of the gravitational lensing of a single source have kept astronomers busy for nearly two decades.

MORE GRAVITATIONAL LENSES

The finding of 0957+561 and the growing likelihood that it reveals a gravitational lens inspired observational searches for more gravitational lens candidates. Before we look at these let us see what theoreticians expect to find, on the basis of Einstein's general theory of relativity.

Figure 6.21(*a*) shows a very symmetrical lens, based on Schwarzschild's solution. We have a source located on the axis joining the observer to the centre of the spherical gravitating mass. In such a case, rays from the source at a particular angle to the axis can leave along any of an infinite number of directions, all lying on a cone with the source as the vertex. They will be similarly bent by the gravitating mass and arrive at the observer along directions lying on another cone. So the observer will see *an infinite number of images all lying on a ring.* This ring is called the *Einstein ring*. In practice, of course, such exact symmetry does not prevail in nature. So we don't expect to see a perfect Einstein ring. Rather, we will see a few images into which the ring may 'break up'. A typical scenario leading to three images is shown in Figure 6.21(*b*).

General mathematical theorems on gravitational lensing by typical astronomical sources suggest that we should normally see an *odd* number of images. Not all these images are of the same intensity, however, as we saw in the case of 0957+561. Thus it may happen that we see only two out of three images, because the third image happens to be very faint. Indeed, there are more cases

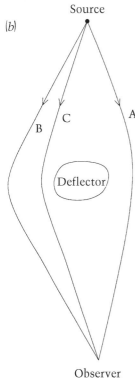

Source

(b)

B   C   A

Deflector

Observer

Figure 6.21: In (a) we see an Einstein ring formed by lensing of a source located on the axis from the observer to the centre of the lensing spherical mass, produced using a lens simulator. In (b) a three-image scenario in the general case of an asymmetric lens is outlined. There are three possible routes, A, B, C, for light rays from the source to the observer. In this case, if the third image is very faint, only two images will be seen.

of an even number of images (two or four) being seen than of an odd number!

Figures 6.22 and 6.23 show two examples of gravitational lens candidates, one containing three images and the other four. In neither case has the lensing galaxy been identified but theoreticians have 'modelled' these cases, with possible suggestions for the mass and distance of a lens galaxy in each case. The maximum angular separation between the images in the triple case is 3.8 seconds of arc.

In some cases it may happen that only *one* image is seen. This would mean that most of the light from the source coming towards the observer is concentrated in only one image, the rest being very faint. The one image seen can then appear exceptionally bright, because of the concentration of light.

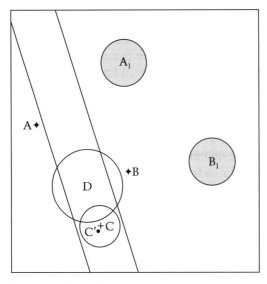

Figure 6.22: The radio source 2016+112 has been considered a good lensing candidate, with three images A, B and C. The optical counterparts of A_1 and B are star-like optical sources with redshifts around 3.27, A being about 30 per cent brighter than B. The image C' was discovered later and is very close to C, which might be an elliptical galaxy. The optical source $C+C'$ is fainter than B by a factor of around 4. The sources A_1 and B_1 have nothing to do with the original quasar. D is an en-route galaxy, with redshift around unity, and may be responsible for the lensing. Models of this system exist but its nature is yet to be fully understood. (From D. P. Schneider *et al.*, **294**, 66, 1985.)

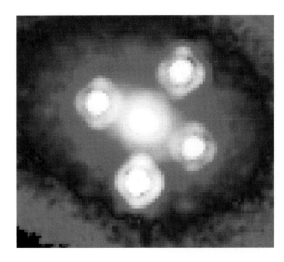

Figure 6.23: Hubble Space Telescope image of the Einstein Cross. Picture by J. Lehar *et al.* from the CFA-Arizona Space Telescope Lens Survey.

The possibility also exists that the lensing mass is altogether invisible, being a massive black hole or a huge lump of dark matter of the kind required by cosmologists (see Chapter 7).

Gravitational lensing not only brightens the image in the above fashion, it can also enlarge the object as a magnifying glass does. Depending on the geometrical arrangement of the source, the lens and the observer, the magnification may be small or large.

Figure 6.24 shows a laboratory-made *simulator* of a Schwarzschild gravitational lens. It is constructed from transparent material, with a radial profile given in Figure 6.24(b). In this case the lens thickness falls sharply away from the centre and then tapers out to the boundary. Light rays passing through it get bent by refraction (as a typical glass lens) but, with this shape, the bending effect is exactly what the gravitational attraction of a spherical mass at the centre would produce.

It is instructive to see how the Einstein ring forms in this simulator only when the source–observer alignment is very symmetric. A breakdown of the symmetry leads to a break-up of the ring into two images.

Arcs and rings

Figure 6.25 shows two examples of arc-like images of galaxies! Are they Einstein rings partially broken up? It is tempting to argue in this fashion but it might not be correct. Let us look at the history briefly.

(a)

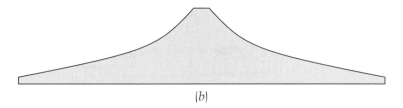

(b)

Figure 6.24: (a) A laboratory-made simulator of a Schwarzschild gravitational lens. (b) The radial profile of the curved surface of the lens in (a). (Model made by P.K. Kunte of the Tata Institute of Fundamental Research, Mumbai, India.)

Figure 6.25: (a) The arc in Abell cluster 370; (b) the arc in cluster Cl 2244. CCD image by Jean-Paul Kneib and Richard Ellis (courtesy of J. P. Kneib and STScI).

In the late 1970s, studies of clusters had given indications of extended arc-like structures but the quality of the data did not permit definitive conclusions. By the mid-1980s, however, the existence of arc-like structures became undeniable, although they were not explicitly sought. During 1986–7, arcs in clusters became well established. Roger Lynds and Vahe Petrosian from the USA and the Toulouse group of G. Soucail, Y. Mellier, B. Fort and J.P. Picat independently reported findings of arcs in clusters.

The arc shown in Figure 6.25(a) is in Abell Cluster 370, and its length is 21 seconds of arc. Its mean thickness is two seconds of arc while the arc radius is 15 seconds of arc. The arc is not uniformly bright and appears knotty. Its redshift has been measured to be 0.724. Using the distance–redshift relation (see Chapter 7), the arc should be about seven billion light years away. Since the cluster itself is much closer to us, the arc does not belong to it.

What was the origin of this arc? Several different interpretations using various astrophysical processes were tried. They did not work. Ultimately the concept of gravitational lensing won out. *We are not seeing a real circular arc in Figure 6.25. We are seeing a distorted image of a galaxy of redshift 0.724, produced by a foreground cluster of galaxies.*

This is something like the image one sees standing before a curved mirror or the image of an extended object viewed through a lens. Modelling of the arc in Abell 370 has now demonstrated how such distorted images form by gravitational lensing.

Other similarly distorted images have been found in other clusters, such as Abell 963 and Abell 2390, all telling the astronomer that what is appearing in the camera is not necessarily what is actually there.

Finally, let us look at an example of a real Einstein ring. When the radio source MG 1131+0456 was studied by the Very Large Array, the contour map shown in Figure 6.26 emerged. The overall shape of the contours is a thick elliptical ring with large axis of size 2.2 seconds of arc and short axis of size 1.6 seconds of arc. There are four other sources (A1, A2, B and C) but no radiation from the interior of the ring.

This type of morphology is very unusual in a radio source and again suggests that we are not seeing reality but a distorted version of it. Theoreticians have modelled this ring with considerable success, assuming that the source itself is extended. Although we do not know the distance of the source, the geometrical details place consider-

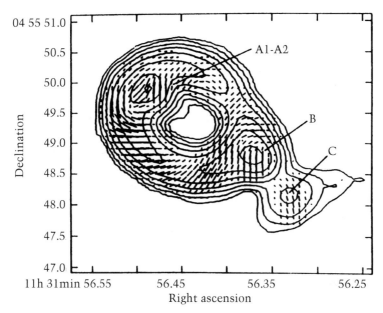

Figure 6.26: The 6-cm wavelength map of the source MG 1131+0456 looks like an Einstein ring (courtesy of J. Hewitt).

able constraints on a lens model, whose success can be judged by how closely it can match the observed images.

BACK TO SUPERLUMINAL MOTION

We return to the observations of superluminal (faster-than-light) motion observed through very-long-baseline Interferometry, which we discussed earlier in this chapter. We mentioned that there were three explanations of this superluminal motion of which we discussed two. We now consider the third hypothesis, which draws on the phenomenon of gravitational lensing. In fact this explanation was proposed by S.M. Chitre and the author in 1976, *three years before* the idea of gravitational lensing shot into prominence.

To understand the explanation, look through a normal lens at two small spheres separated by a short distance (see Figure 6.27). The spheres look farther apart than they actually are. Now imagine that the spheres are slowly moving apart. Viewed through the lens, their

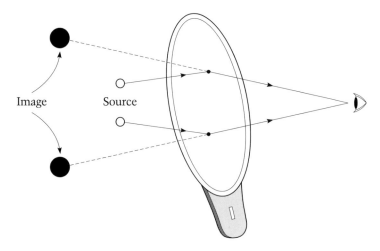

Figure 6.27: The two spheres *A* and *B* when viewed through a lens appear farther apart.

separation, which is magnified, will appear to increase *faster* than the actual rate.

Therein lies the essence of the explanation. Figure 6.28 illustrates the corresponding scenario for a quasar. Imagine that a galaxy lies en route to the quasar, which lenses its two VBLI components. If the galaxy is located at a suitable intermediate distance between these sources and ourselves, we will see the separation between the two VLBI lobes magnified, just as with the hand-held lens of Figure 6.27 a large magnification can be obtained by a lens placed at a suitable distance from the source. And, as the lobes move apart, we should see their speed of separation also magnified.

Calculations with lens models show that a large magnification of speed in this way can make a subluminal speed appear superluminal. The lens idea also has an added bonus. The main image seen with a lens is amplified in brightness, and so picking out a superluminal case is made easier for observers. This circumstance works in favour of finding such cases whereas the somewhat special location of the lensing galaxy makes the phenomenon rather rare. Taking these two conflicting factors together, the lensing scenario turns out to be at least as plausible as the beaming scenario discussed earlier. Further observational studies, such as direct evidence for beaming or the

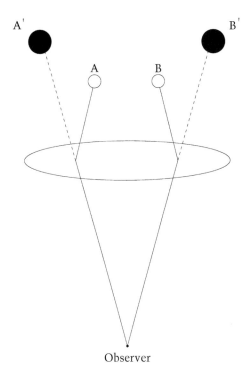

Figure 6.28: An intervening galaxy can act as a lens to magnify the separation between the two VLBI components *A* and *B* of the quasar. The figure shows how the radio waves from *A* and *B* are bent by the galaxy to make them appear to come from *A'* and *B'*. What we actually see are the images *A'* and *B'* moving apart, and their speed of separation may appear to exceed the speed of light even though the actual sources are moving apart with a speed less than the speed of light.

existence of intervening lensing galaxies, will ultimately decide which explanation (if either!) is correct. At present only about 25 or so such sources are known and the statistics are therefore rather poor.

AU REVOIR TO ILLUSIONS

There we leave these wonderful illusions, which may fool the unwary astronomer. We have discussed individual sources and how their appearance may be distorted by gravitational lensing. It may also happen that the entire source population may be distorted by gravitational lensing, leading to errors in the counting and the other survey statistics. This is like measuring the heights of human beings seen through distorting lenses. Astronomers have to make allowances for these effects in interpreting what they see.

However, we have not seen the last of gravitational lensing in this book. We will encounter it again in different contexts in the next chapter, as we move on to look at the largest and grandest wonder of them all, the expanding universe.

The expanding universe ⑦

In this final chapter we will look at the grandest aspect of the cosmos, the large-scale structure of the universe. The universe, by definition, includes *everything* that we can observe physically. The seventh wonder, then, is all about how the universe itself behaves in space and time. How and when did it come into existence? What is its present extent? When, if at all, will it end? Does it contain anything beyond what we can see?

These questions appear philosophical and, indeed, have kept philosophers from different civilizations busy for millennia. A survey of ancient literature shows how humans speculated in the past and found answers to their queries. Where facts were lacking, suitable myths were substituted. But there is no doubt that some of these myths reveal a great maturity of thought.

Today scientists are trying to tackle these issues, in ways based on observed facts coupled with mathematical modelling, even though speculation cannot be kept out altogether. The subject of cosmology is all about these attempts.

It is best, perhaps, to view these modern attempts at understanding against the backdrop of ancient mythologies.

WHAT THE ANCIENTS THOUGHT OF IT ALL

The *Rigveda*, one of the ancient scriptures of the Aryans in India, has this to say:

> At that time (when the universe was not born) there was no 'existence', nor was there 'non-existence'. At that time there was no space nor was there the sky beyond it . . .. There were no means available to

distinguish between night and day . . .. How did the spread of existence come about? Who can tell this in detail? Who knows definitely? Even gods came after the spread of existence. This spread (of existence), where it came from, whether created or not, may or may not be known to the One who presides over the great heavens . . . .

This suggests how scholars in the Vedic times (prior to 1500 BC) raised fundamental questions, answers to which modern cosmologists are still seeking today.

Later on, myths began to take the place of ignorance: answers had to be found to satisfy human curiosity up to a certain level. The mythologies described in the Hindu *puranas* contain several different and highly imaginative ideas.

Figures 7.1 and 7.2 illustrate some of these ideas. The idea of 'Brahmanda', the cosmic egg, involved the concept of the whole universe appearing out of a gigantic egg, while the Earth itself was believed to rest on a hierarchical structure involving four elephants in four directions, who stood on a giant tortoise which was supported by a snake

Figure 7.1: The creation came out of a cosmic egg called *Brahmanda*, according to the ancient Hindu *puranas* of India.

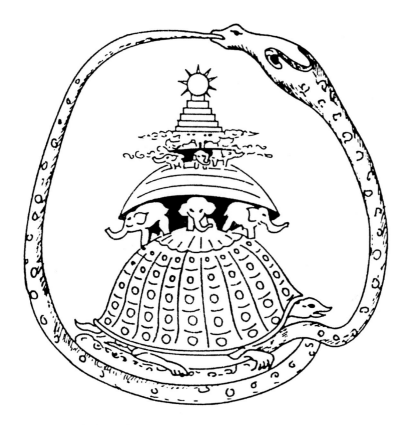

Figure 7.2: The hierarchical model supporting the Earth. The divine serpent *Shesha Naga* carried the entire burden.

swallowing its tail. We will have occasion to refer to Figure 7.2 in a modern context later.

Figure 7.3 shows another concept from a different part of the globe. It is the Norse 'World Tree', which depicts all of the (then) visible universe being carried by the different parts of the tree. Says a Norse myth of circa 1200 AD:

> There was no sand nor sea, nor soothing waves . . .. No Earth anywhere, nor upper heaven . . .. A gaping chasm and grass nowhere. Then the gods Odin and Thor shaped the world. The Earth was flat and in the centre grew the great tree of life, Yggdrasil. The ash tree was watered by three magic springs that never ran dry, and the foliage was always thick and green.

Figure 7.3: The Norse World Tree, which describes the universe in space and time. The three Fates holding the braid of life represent the past, the present and the future.

Coming down to more tangible aspects of the cosmos such as motion in the solar system, the Pythagoreans in Greece, about four centuries before Christ, had a theory that the Earth goes round a central fire (see Figure 7.4), always turning the same face towards it (as does the Moon to Earth, see Chapter 1). The Sun is nowhere within the presumed Earth orbit. When sceptics asked, why don't we see the central fire?, they were told that a 'counter-Earth' always shields it from our sight by moving synchronously with the Earth, as shown in Figure 7.4. The sceptics persisted in asking, why then don't we see the counter-Earth? To which the defenders of the theory replied by saying that Greece was on the other side of the Earth, the side facing away from the counter-Earth. However, the theory was shot down when a few explorers went to the side supposedly facing the central fire and saw neither the fire nor the counter-Earth!

This example was one of the early cases where observation could be used to shoot down a theory. It was the dawn of the scientific

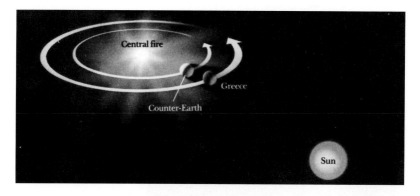

Figure 7.4: The Pythagorean view of the Earth–Sun system (see text for details). From D. Layzer, *Constructing the Universe*, Scientific American Books, 1984. Used with permission of W. H. Freeman and Co.

method where pure speculations would not be accepted but some experimental or observational proof would be required for justifying the theory.

A major step forward in observing the cosmos was taken in 1609 when Galileo Galilei (see Figure 7.5) used a telescope to view the sky. The telescope had been discovered only a few months earlier and its potential in bringing 'far' to 'near' on the Earth had prompted Galileo to adapt the instrument to view the Sun, Moon and planets. The Galilean telescope (see Figure 7.6) was very modest by modern standards, yet it heralded a new era. Indeed, the discoveries made by the telescope, such as the craters on the Moon, the spots on the Sun and four satellites of Jupiter, were all quite unexpected – and unwelcome too.

Unwelcome, because, as it happens with new findings, they can sometimes shake up old faiths. The craters on the Moon (see Figure 7.7) or the sunspots (Figure 7.8) were seen as blemishes on divine creation and, as such, contradicting the tenet that the creation is perfect. Likewise, the belief that everything revolves round the Earth was threatened by the finding that four moons are circling Jupiter (Figure 7.9).

The picture of the universe was slowly built up into its modern form over centuries, with many misconceptions falling by the wayside as the clarity of the picture improved. We will skip these intermediate steps and get to the modern view of the universe.

Figure 7.5:
Galileo Galilei.

AN OVERVIEW OF THE UNIVERSE

The size of the universe can be appreciated in a series of hierarchical structures of increasing size and mass. Figures 7.10 and 7.11 show these steps, the one for linear size and the other for mass. The numbers used here have been rounded up or down from their exact values, just to get an idea of the magnitudes involved.

Starting with the Earth, we know that its radius is 6400 kilometres and its mass is 6000 million million million tonnes. The Sun's radius is about 110 times larger than the Earth's and its mass is more than 300 000 times the Earth's mass.

The Sun is a typical star, and we saw in Chapter 2 that, as stars go, it is average in size, neither large nor small. But there are 100–200 billion stars in our Milky Way Galaxy. Figure 7.12 shows a picture of the Galaxy made up by photographing it in different directions and putting the pictures together. Notice that we are looking at the Galaxy from inside and therefore it is not possible for us to get a full view of it. However, Figure 7.13 shows how the Galaxy would look from different angles. It is disc-shaped and has a bulge at the centre. The disc itself has

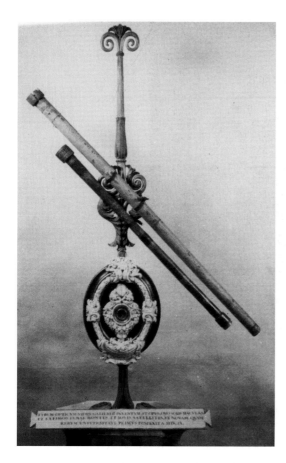

Figure 7.6: Telescopes used by Galileo. The larger telescope had an aperture of 2.6 cm, focal length of 1.33 m; the magnification was ×14.

spiral arms, where the stars are more densely distributed. The location of the Sun and its planetary system is about two-thirds of the way out from the centre of the disc. As seen in the figure, the diameter of the disc is about 100 000 light years.

The next level of structure in this hierarchy is the group in which a galaxy belongs. Our Galaxy is a member of the Local Group, which includes some twenty galaxies. Not all these galaxies are of equal size, however. Our Galaxy and the galaxy Andromeda (catalogue number M31 in the Messier Catalogue) dominate the Local Group. The distance between our Galaxy and Andromeda is about two million light years. See Figure 7.14 for a photograph of this galaxy.

Figure 7.7: A modern picture of the craters on the Moon, first discovered by Galileo with his telescope (courtesy of NASA).

Although these photographs are in the optical, that is, the visible wavelengths, as we have seen there are galaxies that radiate in the infrared, radio or X-rays, in some cases even more than in the optical wavelengths. We saw examples of radio galaxies in Chapter 5.

Figure 7.15 shows a *cluster* of galaxies. A typical cluster may contain hundreds of galaxies. The diameter of a cluster may be in the range 5–10 million light years and it may contain mass equal to several hundred million million solar masses.

For a long time it was believed that the universe does not contain structures larger than clusters of galaxies and that on a scale larger than, say, 30 million light years it is homogeneous. In the last three decades, though, the systematic mapping of galaxies in space and detailed studies of clusters of galaxies have revealed inhomogeneity on a still larger

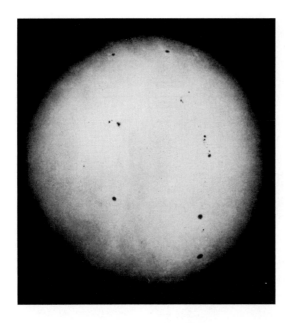

Figure 7.8: Sunspots photographed with modern equipment (courtesy of National Optical Astronomy Observatories).

scale, as shown in Figure 7.16. Here one sees *superclusters* on the scale of 150 million light years with masses of the order of ten to a hundred times that in a cluster. Moreover, these superclusters show a filamentary structure and are separated by *voids*, which also extend to more than 100 million light years.

Does the hierarchy extend to an even higher level? For the present there is no indication of this; but it would be fair to say that astronomers have not yet been able to analyse systematically regions of size, say, 500 million light years, to see whether there is clumpiness on such scales.

The very highest scale of length in Figure 7.10 is that of the universe itself! Actually, the universe may be boundless, but the distance out to which we can probe with our best telescopes is of about 10 000 million light years. And the mass contained in a sphere of this size may be several thousand million million million solar masses, as shown in Figure 7.11.

When we review this complex and gigantic structure we really begin to appreciate the smallness of our terrestrial environment. We live on a tiny planet that goes round a star which is a member of a galaxy containing a hundred thousand million similar stars, a galaxy that is a member of a smallish group which is part of a cluster that belongs to a supercluster, which in turn is one of many such in a vast, maybe boundless,

Figure 7.9: The four inner satellites of Jupiter, first seen by Galileo. Jupiter is known to have at least 16 satellites. This picture was taken by the Voyager 1 spacecraft in March 1979 (courtesy of NASA).

universe. In a strange way this hierarchical set-up is reminiscent of the Hindu puranic hierarchy described earlier.

Eddington has described the daunting challenge facing the cosmologist in the following words:

> Man in search for knowledge of the universe is like a potato bug in a potato in a sack lying in the hold of a ship, trying to discover, from the ship's motion, the nature of the vast sea.

Yet cosmologists have taken up the challenge and it is to their credit that they have made significant progress in piecing together at least a

(Linear Size)

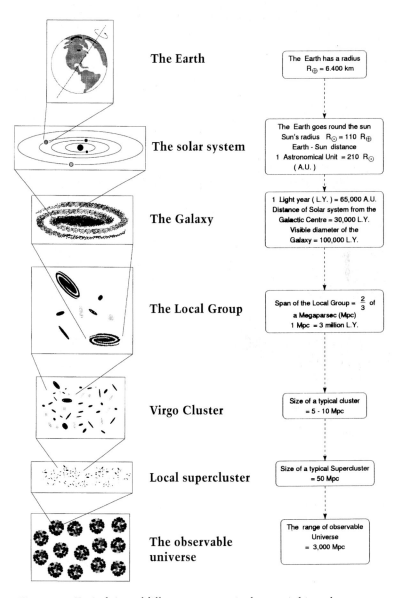

The Earth		The Earth has a radius $R_\oplus$ = 6.400 km
The solar system		The Earth goes round the sun Sun's radius $R_\odot$ = 110 $R_\oplus$ Earth - Sun distance 1 Astronomical Unit = 210 $R_\odot$ (A.U.)
The Galaxy		1 Light year (L.Y.) = 65,000 A.U. Distance of Solar system from the Galactic Centre = 30,000 L.Y. Visible diameter of the Galaxy = 100,000 L.Y.
The Local Group		Span of the Local Group = $\frac{2}{3}$ of a Megaparsec (Mpc) 1 Mpc = 3 million L.Y.
Virgo Cluster		Size of a typical cluster = 5 - 10 Mpc
Local supercluster		Size of a typical Supercluster = 50 Mpc
The observable universe		The range of observable Universe = 3,000 Mpc

Figure 7.10: Typical sizes of different structures in the cosmic hierarchy.

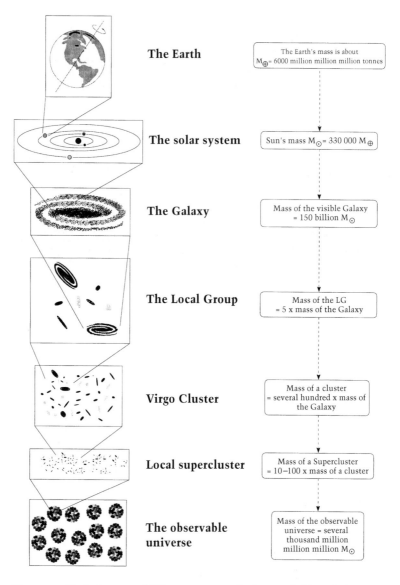

The Earth

The Earth's mass is about
$M_\oplus$ = 6000 million million million tonnes

The solar system

Sun's mass $M_\odot$ = 330 000 $M_\oplus$

The Galaxy

Mass of the visible Galaxy
= 150 billion $M_\odot$

The Local Group

Mass of the LG
= 5 x mass of the Galaxy

Virgo Cluster

Mass of a cluster
= several hundred x mass of
the Galaxy

Local supercluster

Mass of a Supercluster
= 10–100 x mass of a cluster

**The observable
universe**

Mass of the observable
universe = several
thousand million
million million $M_\odot$

Figure 7.11: Typical masses of different structures in the cosmic hierarchy.

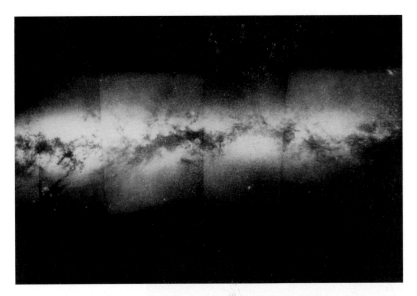

Figure 7.12: A composite photograph of the Milky Way obtained by putting together photographs in different directions (courtesy of Palomar Observatory, California Institute of Technology).

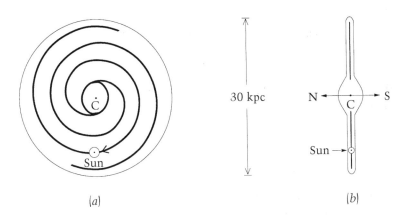

Figure 7.13: The Milky Way seen (a) face on and (b) edge on. The unit of length used is the kiloparsec (kpc) which is approximately equal to three and a quarter thousand light years. The diameter of the Galaxy is 30 kpc or 100 000 light years.

Figure 7.14: The Andromeda galaxy (photograph by the National Optical Astronomy Observatories).

partial understanding of the puzzle of the universe. As Albert Einstein once put it:

> The most incomprehensible thing about the universe is that it is comprehensible.

This is our seventh wonder, the universe itself, with all its remarkable properties revealed so far and all its tantalizing secrets still to be discovered.

WHY IS THE SKY DARK AT NIGHT?

We begin with this simple question, which hardly seems relevant to cosmology. It is part of a purely daily experience, telling us that the Earth spins about its axis with a period of 24 hours and the part of its surface facing away from the Sun experiences darkness which is nightfall. Is this not sufficient answer to the question?

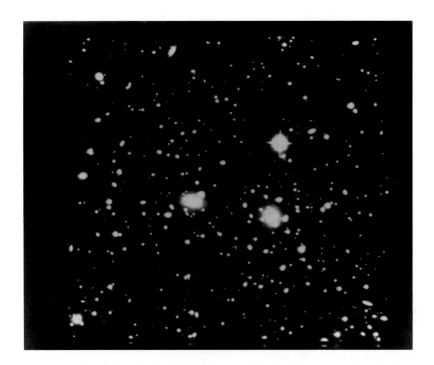

Figure 7.15: The Coma cluster of galaxies (photograph by the National Optical Astronomy Observatories).

Heinrich Olbers, a German astronomer, was not satisified with this answer. In 1826 he did a calculation whose answer was so startling that it kept astronomers busy for a century and a half trying to find where Olbers had gone wrong. For if he were right, then the sky should be extremely bright all the time, irrespective of which side of the Earth the Sun is on.

Known as the *Olbers paradox*, the argument is essentially as follows.

Besides the Sun, the sky contains very many other stars which are also emitting light, some of which will come towards us. Of course, the light from a typical star will be quite minuscule as the star is very far away. However, Olbers argued that there are so many stars in the universe that their combined contribution might not be negligible. So he set out to compute it using a simple argument.

Imagine that the universe is infinite in extent and is uniformly filled with stars, all of them like the Sun. Suppose we draw a sphere of radius R and consider a thin shell on its surface (see Figure 7.17). The surface

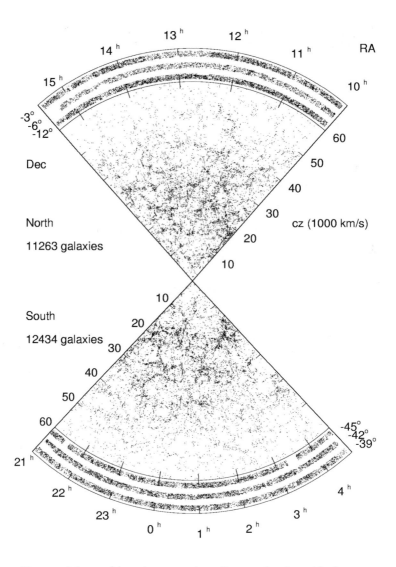

Figure 7.16: A map of the universe up to large distances showing voids, clusters and filaments in the distribution of galaxies, which appear as dots in the picture. The figure describes the Las Companas Redshift Survey, which has some 26 418 galaxies with typical redshifts of the order of 0.1, spread over about 700 square degrees of the sky. The redshifts are translated into distance using Hubble's law. (From H. Lin *et al.*, *Astrophysical Journal*, **471**, 617, 1996.)

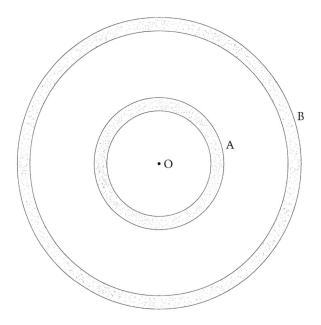

Figure 7.17: Stars in a typical shell around the observer O at the centre contribute a radiation flux at O that does not depend on the distance of the shell. Thus two shells of the same thickness, A and B in the figure, contribute the same flux at O.

area of the sphere is $4\pi R^2$, and if the shell has thickness a, its volume will be approximately the area multiplied by thickness, that is, $4\pi R^2 a$ (we have imagined opening out the spherical shell as a flat sheet). Further, if the universe has N stars per unit volume, then the number of stars in this shell will be $4\pi R^2 aN$. Now imagine a typical star in this shell as having luminosity L. Then the amount of its radiation passing through unit area at the centre, O, will be $L/(4\pi R^2)$. (Issues such as the radiation received from a star are discussed in greater detail in Chapter 2.) So we see, on multiplying this quantity by the number of stars in our shell, that these stars contribute a total flux of radiation equal to LNa at our location. Notice that the result *does not depend on the distance of the radiating stars.*

The last part of Olbers's argument is now straightforward. For any observer, such as us, divide the entire universe into concentric spherical shells all of the same thickness. As we have shown above, each shell contributes the same flux at the observer. *But the number of such shells is obviously infinite.* Therefore it follows that the total flux from *all* stars in the universe is also infinite!

This was the logical conclusion that Olbers came to with his basic assumptions. Now we see why whether or not we are facing the Sun, the night sky should be infinitely bright.

But the night sky is dark. So there is something wrong with the above estimate. But where is the mistake?

Careful consideration of all Olbers' arguments shows one loop-hole. The stars are not point sources: they have a finite size. So when we start putting stars in successive shells around O, a stage will come when they will naturally fill the entire sky visible to O. An analogy may help here. If you look through a gap in the trees in a park you can see buildings in the background. However, if you are in a forest of trees, you simply can't see beyond a limited distance. All gaps in the foreground trees are eventually covered by rows of trees in the background. So, even if we draw an infinite number of shells to fill up the whole universe, only the stars in the relatively nearby shells will contribute to the total radiation flux. The flux is therefore not infinite but finite.

But we are not out of the woods yet! For this finite total radiation flux can be computed and it turns out to be as high as at the surface of the Sun. This means that the sky should not only be bright but its temperature everywhere, including in our neighbourhood, should be in the region of about 5500 degrees Celsius. Again we have arrived at an impossible conclusion.

Astronomers in the past suggested two other resolutions of the paradox. The first is that the universe may not be infinite in extent, as Olbers assumed, but finite. This would mean that when we draw our spherical shells we stop at a certain distance beyond which nothing exists. This distance would have to be at least as large as the range of our best telescopes. For, so far as we can see, there is no end to the sources of light up to the distance of some ten billion light years that we can presently probe. In such a situation we do get a resolution of the paradox, for the contribution of sources out to this distance is actually negligible compared to the light we get from the Sun.

The other resolution was that the stars that we see or can in principle see came into existence a finite time ago. Suppose the universe itself came into existence ten billion years ago. Then today we can receive light from only those stars that lie within a distance of ten billion light years. For stars that exist beyond this limit, the light has not had time to reach us yet. Figure 7.18 illustrates this scenario.

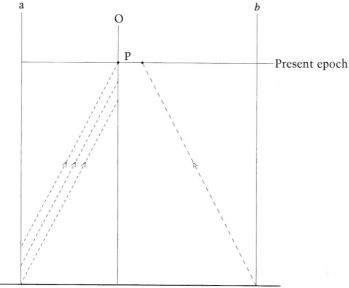

Epoch of origin of the universe

Figure 7.18: In this spacetime diagram, we see the worldlines of two light sources *a, b* located at distances 8 and 12 billion light years from the observer *O*. The broken lines show light tracks. Notice that light signals sent by *a* after the epoch of origin of the universe can reach *O* before the present epoch at *P*, but a signal from *b* has not yet reached *O*.

A further possible resolution of the paradox takes note of the fact that stars in any shell will last for a finite time. They cannot go on shining for ever. We saw in Chapter 2 how any star eventually reaches the end of its energy reservoir. So we cannot expect to find shining stars in all shells for ever. This also effectively reduces the net contribution to the total flux received.

All these arguments have some unsatisfactory feature. For example, if all stars last for a finite duration then, in an infinitely old universe, no shining stars should be left, unless fresh stars are continually being made. A universe which came into existence a finite time ago also raises philosophical and conceptual questions, as does the idea of a universe of finite extent.

Nevertheless, a crucial element had been ignored in Olbers' calculation, and it came to light only in the middle of this century, when Hermann Bondi discussed this issue in a revival of the Olbers paradox.

We will now look at that crucial piece of evidence about the real universe that was not available to Olbers. It is this evidence on which modern cosmology is founded.

HUBBLE'S LAW

Modern observational cosmology was really launched by the discovery of Edwin Hubble (Figure 7.19), announced in 1929, in a paper entitled 'A relation between distance and radial velocity among extragalactic nebulae' in the Proceedings of the National Academy of Sciences of the USA. What Hubble found was a culmination of several years' work on

Figure 7.19: Edwin Hubble, standing in front of the Palomar Schmidt Telescope, with 48 inch diameter mirror. This telescope was ready just before Hubble's death. (Photograph courtesy of Mount Wilson and Las Campanas Observatories.)

the spectra of galaxies, which had started with V.M. Slipher in 1914. To understand its significance, let us look at Figure 7.20.

In this figure, we have photographs of several galaxies in clusters, given one below the other on the left. As we go down the list, the galaxies get fainter and smaller, an indication that we are looking at ever more distant galaxies. The figures giving the actual distances, ranging from 78 million to 3960 million light years, confirm this expectation.

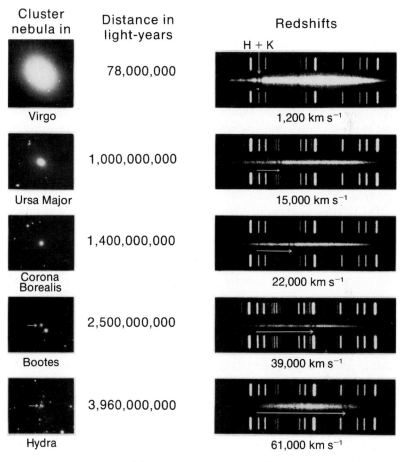

Cluster nebula in	Distance in light-years	Redshifts
Virgo	78,000,000	1,200 km s^{-1}
Ursa Major	1,000,000,000	15,000 km s^{-1}
Corona Borealis	1,400,000,000	22,000 km s^{-1}
Bootes	2,500,000,000	39,000 km s^{-1}
Hydra	3,960,000,000	61,000 km s^{-1}

Figure 7.20: In each pair of photographs above we see a galaxy in a cluster on the left and its spectrum on the right. The distance of the galaxy is given along with its radial velocity (under each spectrum) estimated according to the Doppler effect (Palomar Observatory, California Institute of Technology).

When discussing stars in Chapter 2, we indicated how the observed faintness of a star may be used to estimate its distance. The same rule applies to galaxies. Assuming that galaxies do not vary too much in their intrinsic luminosity, we expect a fainter galaxy to be more remote than a brighter one and can use the quantitative measurement of its faintness to estimate its distance. Likewise, on the grounds that if all objects of a given class are of the same size then the farther one will look smaller, we can double-check on the estimated distance.

As we will see later, both these reasonable-looking assumptions can lead us into pitfalls. For the time being, however, we will assume their correctness.

Looking to the right we see the spectrum of each galaxy. The actual spectrum is in the centre with a comparison spectrum on each border. The latter is a spectrum of a laboratory source showing the dark (absorption) lines. The actual spectrum also has one or two dark lines, which are shifted towards the red end (longer wavelengths) relative to those in the comparison spectrum. That is, we are witnessing an example of *redshift*, which we encountered earlier in Chapter 5. If we interpret it as a case of the Doppler effect, then we can compute the speed of the galaxy away from us. This is the speed given under each spectrum.

The relation used to estimate this effect is very simple to understand. The 'redshift' of a spectral line is measured in terms of the fractional extent by which its wavelength has increased relative to that in the comparison spectrum. Thus, if the line normally has a wavelength of 500 nanometres,[14] but appears in the spectrum with a wavelength of 505 nanometres, the shift is of five nanometres. As a fraction of its original wavelength this shift is 5/500, that is one per cent. This is the redshift of the line.

How is this information used to estimate the speed of recession? This is where the Doppler effect comes in. The rule given by this effect is simple: *multiply the redshift by the speed of light*. So, in the above example, the speed of recession will be one per cent of the speed of light, that is, 3000 kilometres per second.

Although Figure 7.20 does not show the very first data reported by Hubble in his 1929 paper, it gives an idea of what he had found. His finding was indeed remarkable; for even at a glance we can see that the

---

[14]A nanometre is a billionth part of a metre (10^{-9}m).

farther galaxies are moving faster away from us. Hubble found a more precise relation which can be stated thus:

The speed of recession of a galaxy relative to us is proportional to its distance from us.

In short, if we have two galaxies, G_1 and G_2, the latter being twice as far from us as the former, then the speed of recession of G_2 will be double that of G_1.

This result continued to hold when later extended to more and more remote galaxies. It became known as *Hubble's law*. It tells us that the velocity of recession of a galaxy is obtained by multiplying its distance by a fixed constant, called the *Hubble constant*. Hubble had estimated that a galaxy at a distance, of, say, ten million light years, would be moving away with a speed of about 1600 kilometres per second. Figure 7.21 demonstrates how this relation between velocity (radially away from us) and distance looks when plotted on a graph. As we shall see later, however, Hubble had grossly overestimated the value of this constant.

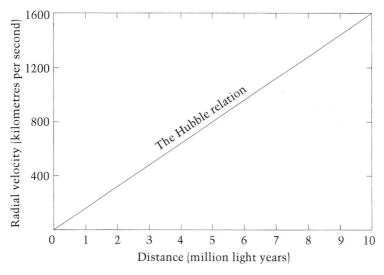

Figure 7.21: In the above graph the velocity of a galaxy (estimated from its redshift) and its distance from us are plotted respectively on the vertical and the horizontal axes. The points fall on a straight line, whose slope on the horizontal axis determines the value of Hubble's constant. The value used here is that originally quoted by Hubble, which is now known to be too high.

Taken at its face value, the following result emerged from Hubble's law. We see galaxies receding from us in whichever direction we look, and the farther galaxies are moving faster. Does that place us, that is, our Milky Way Galaxy, in a special position in the universe? We cannot resist giving a historical perspective while answering this question.

In the olden days, dating back thousands of years, the general belief was that the Earth is at rest at the centre of the cosmos, whereas the whole celestial firmament revolves round it. The special importance enjoyed by the Earth lasted until the sixteenth century, when the work initiated by Nicolaus Copernicus established the Sun as the centrepiece of the planetary system. Two centuries later, William Herschel prepared a map of our Galaxy based on his studies of stars and their estimated distances. In this map he placed the Sun at the centre of the Galaxy. Herschel's map is shown in Figure 7.22.

So, even though the Earth had been dislodged from a privileged position, we could still boast of a special position for our Sun and the planetary system. But this special status enjoyed by the Sun vanished when in the early part of this century the picture of our Galaxy given in Figure 7.13 because established. Harlow Shapley at the Harvard College Observatory was responsible for bringing about the correct perception, according to which the Sun is far from the Galactic centre. The current estimate of its distance from the latter is about 30 000 light years.

Having given up the special status of the Sun in our Galaxy, the anthropocentric view shifted to another level. Is our Galaxy the most important object in the universe? The answer even at the turn of this century was in the affirmative.

Immanuel Kant (1724–1804) had proposed a contrary view two centuries earlier, namely that there exist other systems of stars in the

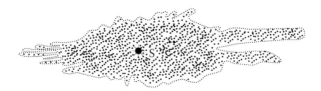

Figure 7.22: In 1785 William Herschel drew this map of our Galaxy. Notice that the star marked at the centre is the Sun.

universe similar to our Milky Way, only they are so far away from us that we cannot see them in detail. He called them *island universes*.

Kant's view was, however, much ahead of its times and found very few takers. In Figures 7.23–7.25 we see some nebulae, that is, cloud-like images which are luminous but not concentrated sources of light like stars. Today we know that the nebulae of Figures 7.23 and 7.24 are within our Milky Way, while the nebula in Figure 7.25, as well as the Andromeda Nebula of Figure 7.14, are actually external galaxies in their own right.

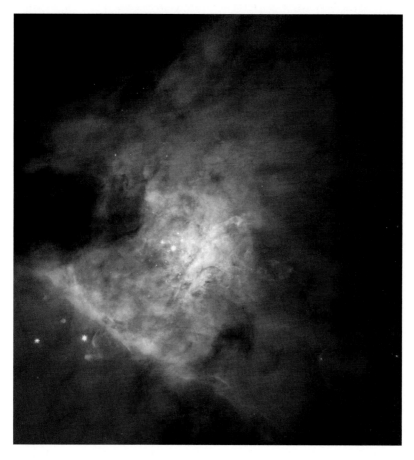

Figure 7.23: Mosaic of 45 separate images of the Orion Nebula put together to describe this vast cloud-like region in the Galaxy (photograph by C.R. O'Dell of Rice University and NASA and STScI).

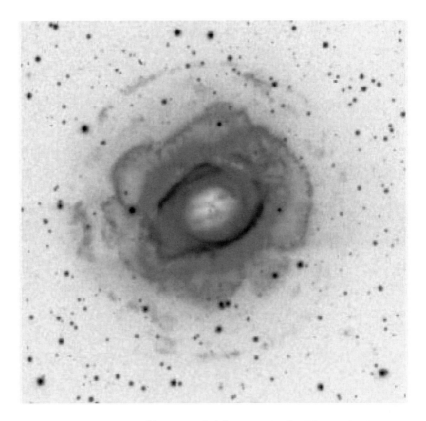

Figure 7.24: CCD image of the Ring Nebula by George Jacoby. Whereas the picture shown in Figure 3.11 is like the traditional photographs of the nebula, this image shows considerably greater detail (courtesy of G. Jacoby).

However, even during the eighteenth and nineteenth centuries, controversy centred round the distances of some of these nebulae, especially those that do not seem to lie in the disc of the Milky Way. Like Kant, the mathematician Johann Lambert (1728–77) had argued that some of these nebulae are extragalactic, being galaxies in their own right. R.A. Proctor (1837–88) offered an explanation in support of the Kant–Lambert ideas. He suggested that the reason why similar nebulae were not found in the disc of the Milky Way was the presence of dust, which absorbs the light passing through it. The dust is concentrated in the disc and it blocks light passing in the plane of the disc, whereas there is not much blockage of light travelling perpendicular to the disc. This explanation eventually turned out to be true. Nevertheless, even in

Figure 7.25: Galaxy in Sculptor, NGC 253 (Palomar Observatory/California Institute of Technology).

1919, the same Harlow Shapley who had correctly placed the sun in the Galaxy had this to say:

> Observation and discussion of radial velocities, internal motions, and distribution of the spiral nebulae, of the real and apparent brightness of novae, of the maximum luminosity of galactic and cluster stars, and finally of the dimensions of our galactic system, all seem definitely to oppose the 'Island Universe' hypothesis of the spiral nebulae . . .

Again, the correct picture emerged shortly afterwards when, thanks to the availability of the 100-inch telescope at Mount Wilson in 1917, Hubble was able to confirm the Kantian hypothesis and establish the extragalactic nature of spiral nebulae such as the one in Figure 7.14. Thus our Galaxy lost its unique and paramount status in the universe!

It is against this background that we now look at Hubble's own finding, which seemed again to place our Galaxy in a special position – from which all other galaxies are receding. However, this return to glory was short lived. It soon became clear from the mathematical nature of Hubble's law that it treated all galaxies uniformly. Thus, if we perform a thought experiment, place ourselves on another galaxy and observe the universe from there, we will find the same situation: all other galaxies are receding from our new vantage point. Figure 7.26 demonstrates how this happens.

Indeed, the correct way to look at the situation is to imagine the entire space in which the galaxies are embedded as expanding. A gastronomical analogy would be the baking of a cake containing nuts. As the dough bakes, it expands and the embedded nuts move away from one another.

Therefore a natural deduction from Hubble's law is that *the universe is expanding.*

The redshift–distance relation

This was indeed a remarkable conclusion to arrive at, showing that not only is the universe dynamic on the largest scale, but its motion has a very definite pattern. Although Hubble's observations in the 1920s and 1930s were limited to distances hardly more than 100 million light years, later astronomers sought to extend the survey of galaxies to larger distances. And, since Hubble's time, the world's best telescopes have been put to the task of checking whether the law holds at larger and larger redshifts. That is, one would like to check whether galaxies with larger redshifts are fainter, as Hubble found for nearby galaxies. Figure 7.27 shows how the redshift–faintness relation looks for a special class

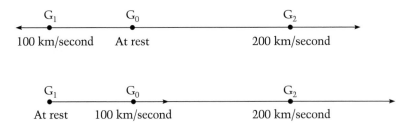

Figure 7.26: Imagine that we are observing two galaxies G_1 and G_2 on opposite sides of our location, G_0. Assume that G_2 is twice as far from us as G_1. Suppose that galaxy G_1 is moving away from us with speed 100 km per second. Then, according to Hubble's law, G_2 will be moving away as shown with a speed 200 km per second. If we now go to G_1 and observe from there, we have to correct for its motion relative to G_0. So from G_1 we will see G_0 moving away with speed 100 km per second and G_2 with speed 300 km per second, as shown in the lower line of the figure. But from G_1, G_2 is three times farther than G_0. So Hubble's law holds at the new observation post also.

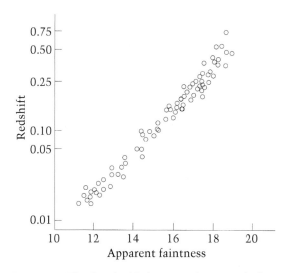

Figure 7.27: The plot of redshifts against faintness of galaxies shows an almost straight-line curve when the sample of galaxies is made up of the brightest members of their respective clusters (based on the work of Allan Sandage and collaborators).

of galaxies. These galaxies are the first-rank members of their clusters, that is, they are the brightest in their respective clusters. Why choose such a special class?

Recall from Chapter 2 that if all sources of light of a given type, like stars or galaxies, were looked at from different distances, those that look fainter will be farther away. This is a basic tenet on which the astronomer bases distance estimates. But there is a pitfall here. If object *A* is intrinsically less powerful a radiator of energy than object *B*, then from the same distance we would find *A* looking fainter than *B*, and would wrongly conclude that it is farther than *B*. To arrive at a correct comparison of distances, we must ensure that all our sources are *equally powerful*.

Allan Sandage, who had been a research student of Hubble, has played a leading role in extending the Hubble relation to greater and greater distances. Using clusters at known distances, he found that the first-ranked member galaxy in any cluster has a luminosity that is not very different from the luminosity of a similar galaxy in another cluster. Thus if we choose such galaxies from different clusters, then the observed faintness of a galaxy will gives us a reliable estimate of its distance. Figure 7.27 bears out this expectation because we find very little scatter around the line drawn through the points.

This is the reason that astronomers have come to regard redshift as an indicator of the distance of an extragalactic object such as a galaxy or quasar. The rule of thumb is that you multiply the redshift by a fixed distance scale to get an estimate of the distance of the object. That fixed distance scale is in turn determined by the value of Hubble's constant. What is that distance scale?

We have referred to Hubble's own estimate of this distance scale as being wrong. There were several systematic errors in the early measurements which were largely responsible for this. As the nature of the errors became better understood, the value of Hubble's constant steadily came down over the years. How does it compare with Hubble's value as in Figure 7.21? Today's values would place a galaxy moving away with speed 1600 km per second at a distance somewhere between 60 to 85 million light years, as opposed to Hubble's estimate of around 10 million light years.

Unfortunately, despite nearly seven decades since Hubble's original paper, astronomers have not been able to fix the value of Hubble's constant within reliable limits, that is, within limits of no more than ten per cent error. There are several 'if's and 'but's in its measurement, with the result that observers have not yet been able to agree a value that can be considered as the 'true' value of the Hubble constant.

If we divide the speed of light by Hubble's constant we get a distance scale. Because of the uncertainty in the magnitude of the Hubble constant, this distance scale is also uncertain to a degree. We shall use here a value equal to ten billion light years, only to fix ideas. It is a round number which conveys the enormity of the cosmic distance scale. As mentioned earlier, we get an idea of how far a galaxy is by multiplying this distance scale by the redshift of the galaxy.

The Olbers paradox revisited

We now return to the innocuous question asked by Olbers: Why is the sky dark at night? For we have now a new element of information about the universe that Olbers and his contemporaries lacked. We know that the universe is expanding and light from any extragalactic source is redshifted.

The redshift works in two different ways to reduce the contribution of the more distant sources to the local radiation background. First, we recall from Chapter 5 that the redshift indicates a comparison of the rates of flow of time at the source and the observer. Although that was noticed in the context of gravitational redshift, the effect is true of any redshift, such as the one found by Hubble. If the redshift is 0.5, then the clock at the observer will run 1.5 times faster compared to the clock at the source. That is, a one-second interval at the source corresponds to a 1.5-second interval at the observer, if we set up a signalling arrangement from the source to the observer. Therefore the rate at which the radiation is received by the observer will need to be *reduced* by a factor 2/3 when estimating the contribution of the source.

Second, the radiation itself is downgraded in energy as it travels through the expanding universe. It is made up of light quanta called photons, each photon having energy proportionate to its frequency. Redshift reduces the frequency of the photon by the time it reaches the observer, and therefore the observer receives a less energetic quantum. In the above example, only two-thirds of the energy per photon radiated by the source is received by the observer.

When we consider these two effects together we find that the farther sources of light contribute much less to the sky brightness than Olbers had estimated. At a redshift of 0.5 the reduction is by a factor 4/9, at a redshift of 1 it is by a factor 1/4 while at a redshift of 9 it will be as low as one per cent. The farther the source is, the greater is the loss in its contribution to the radiation background near us. This is why the total amount of radiation received at an observer is negligible.

Thus a major reason why sky remains dark at night is that the universe is expanding!

THE BIG BANG MODELS

The Olbers paradox shows how a relatively simple-minded question leads to profound cosmological concepts such as the expanding universe. To most people, the concept of the expanding universe is indeed awe-inspiring. There are several questions that occur when coming to grips with this remarkable finding. What is the universe expanding into? What lies outside it? Will the expansion continue for ever? Will it come to a halt and switch over to contraction? If the universe has been expanding and was smaller in size in the past, was there an epoch when it was much smaller, even *point-like with zero volume*? Did it originate in that state? If so what was there before?

Recall the questions asked by the Vedic scholars more than three millennia ago and mentioned at the beginning of this chapter. They find echoes in modern questions about the universe.

To answer them, however, today's cosmologist takes help from the established laws of science, in particular those that may have to do with the very largest-scale structures in the universe. From our previous discussions in this book we see that the most relevant interaction here is gravitation. It is most effective whenever there are large masses and it is all pervasive. We have also seen that Einstein's interpretation of gravitation is more appropriate than Newton's.

Indeed, even before Hubble announced his result, general relativists were busy making models of the universe. Einstein himself had made a beginning in 1917 by proposing a static, homogeneous and isotropic model of a universe. By 'homogeneous', we mean that the universe looks the same at all points in space. By 'isotropic', we mean that the universe looks the same in all directions. In other words, if you are taken to any part of the universe you cannot find any local landmarks to tell you where you are or any local direction to tell you which way you are looking. These simplifying assumptions helped in solving the difficult equations of relativity. Fortunately, the universe does look homogeneous and isotropic on a large enough scale.

The Einstein universe has another property: *it is closed*. That is, if you shine a torch and send out a light ray, it will travel all the way round the universe by gravitational bending and come back to the original spot from behind! Figure 7.28 illustrates this geometry. A closed universe has a finite volume, but no boundary.

The Einstein model, however, lost popularity once it became known that the universe is not static but expanding. Cosmologists then turned to models which described an expanding universe. A few years before Hubble's discovery, several theoreticians had proposed such models, which were initially looked at as mere mathematical curiosities. Models by the Russian cosmologist Alexander Friedmann, the Belgian Abbé Lemaitre and the American H.P. Robertson thus became the starting points for describing cosmology.

The simplest of these models include three types. A type I universe has zero-curvature space, a type II universe has positive-curvature space while the space in a type III universe has negative curvature. (See Chapter 5 for a discussion of the curvature of space.) The dynamical behaviour of all these models has one common feature. They tell us that the expansion of the universe was no slower in the past than it is today.

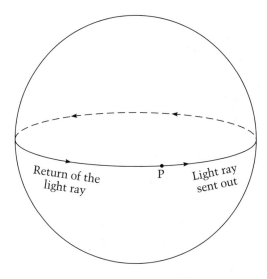

Figure 7.28: Imagine a two-dimensional universe which is confined to the surface of a sphere. The surface has a finite area but no boundary. A flat creature could slide along it endlessly without encountering an edge or a boundary. A ray of light sent to the right from point P will describe a great circle and return to P from the left. The Einstein universe was a three-dimensional version of this picture.

The universe therefore had a smaller and smaller size in the past, being of zero size at a special epoch. This epoch is called the *big bang* epoch, and at that epoch the universe was exploding with infinite velocity, and was in a state of infinite density and temperature. The expansion we see today is the remnant of that gigantic explosion. Are there any other tangible relics? We will come to this question in due course.

What was there *before* the big bang? Nothing! Not even space and time. In fact the state of the universe at this epoch was so peculiar that it defies a physical description. One can start the cosmic clock ticking just after this primordial event brought the universe into existence. In fact it is very similar to the singular state described as the end point of an object undergoing gravitational collapse (see Chapter 5). As mentioned there, the difference is that the universe is *exploding* from that state, rather than *imploding* into it.

We can measure the 'age of the universe' on this cosmic clock as the time elapsed since the big bang. This value is uncertain to the extent that we do not know the true value of Hubble's constant. For the type I model the age works out in the range of 8–10 billion years. For type II models it is lower, while for type III models, it is somewhat higher. Notice, however, that the day and night of Brahma as quoted in ancient Hindu scriptures (see Chapter 5), at 8.64 billion years, came pretty close to this value!

Is the universe open or closed?

The models described above share a common history but not a common future. In particular, all models of type III describe a universe which goes on expanding for ever. In such a universe, our Galaxy will ultimately be left with no neighbours, all having dispersed to infinite distances. A very lonely state indeed! But perhaps preferable to what would happen to us if we were part of a type II universe. Such a universe would go on expanding for a while but would eventually slow down to a temporary halt and begin to contract. It would continue contracting and getting smaller until it reached a state of infinite density and infinite temperature. Exactly the opposite of the big bang, this state is called the *big crunch*: everything in the universe would be crushed to a state of infinite density. Figure 7.29 shows an artist's conception of the big bang, while Figure 7.30 illustrates these two possible future states.

What about the type I model? It is unique of its kind and stands on the dividing border of the type II and type III models. Thus it too expands for ever, like type III models, but only just; a slight decrease in its expansion velocity and it would eventually shrink like a type II model. The type I model is also known as the Einstein–de Sitter model, since it was advocated by Einstein and de Sitter in a joint paper in 1932.

As we mentioned earlier (see Chapter 5), Einstein's theory of graviation relates the geometry of spacetime to the state of its contents. Thus the future of a model depends on its geometry. *All closed expanding models would eventually contract to hit the big crunch. Likewise, all open models would disperse their galaxies to infinity.*

For the simplest models, which we are considering here, there is a rather neat way of distinguishing between the open and closed varieties. This rule of thumb is as follows. *Measure the density of matter in the universe. If it exceeds a certain critical value, then the universe is closed. If not, it is open.*

What is the critical value? Einstein's equations tell us that this is the density one would expect in a type I universe. Theoreticians can determine it if they know the correct value of Hubble's constant. Since we have already found that the true value of Hubble's constant is not easy to fix and that its measurement continues to be embroiled in controversy, we do not know the value of this critical density exactly. Again we shall use a round value, as indicative only: it is a tiny fraction of the density of water, a fraction as small as ten parts in a million million million million million. We shall refer to it as the *critical density* or the

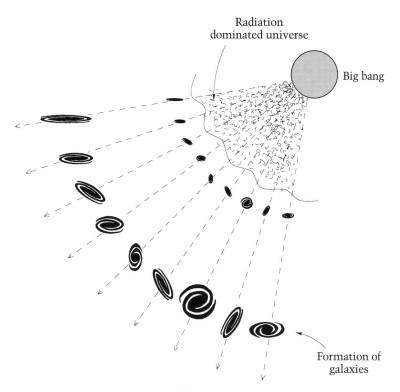

Radiation
dominated universe

Big bang

Formation of
galaxies

Figure 7.29: An artist's conception of the big bang. The technical description of
this state is highly bizarre.

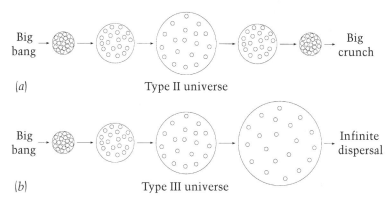

Big
bang

Big
crunch

(a)　　　　　　　　　Type II universe

Big
bang

Infinite
dispersal

(b)　　　　　　　　　Type III universe

Figure 7.30: In (a) we see the eventual future state of a universe of type II: all
galaxies are crunched together as the universe shrinks rapidly to a state of infinite
density. In (b) we see the future of a type III model, wherein a typical galaxy is left
with no neighbours, all having dispersed to infinite distances.

closure density. The latter appellation conveys the fact that in order to 'close the universe', the density of matter must exceed this value.

The density of the universe is not, however, easy to determine. There are complications, which we will discuss shortly, that prevent a straightforward answer. In fact these complications have once again underscored the maxim *'seeing is not believing'*.

An indirect way of testing whether the universe is open or closed makes use of the effect of matter on light. This test was proposed by Fred Hoyle in 1958 and it brings out a remarkable feature of non-Euclidean geometry.

CAN DISTANT OBJECTS LOOK BIGGER?

Let us first review an everyday observation. When we look out to greater and greater distances we see objects getting smaller and smaller. A two-storey building very close to us may dwarf a twenty-storey skyscraper two blocks away. A hiker in the mountains may think that the distant peaks do not look particularly tall, only to discover on getting closer that they are very tall indeed.

There is a simple explanation for this effect. The perception of the size of an object is based on the angle subtended by the object at the eye (see Figure 7.31). And the more distant the object the smaller is this angle in inverse proportion to the distance. For example, suppose we view a tree from a distance of fifty metres and then from five hundred metres. Its size will appear smaller by a factor ten in the second case compared to the first.

This result, however, depends on Euclid's geometry. The same may not hold, for example in an expanding universe, wherein the geometry of spacetime is non-Euclidean.[15] This is what Hoyle pointed out. As we have seen, the track of a light ray through space depends on the spacetime geometry. We have also seen how a light path can be bent by enroute matter and can lead to gravitational lensing. Thus we do expect a departure from the Euclidean behaviour described above. And the more matter in the universe, the greater will be the departure.

Let us now look at the result Hoyle derived for the expanding universe model. Figure 7.32 contrasts the behaviour of a population of

[15]See Chapter 5 for a general discussion of non-Euclidean geometry.

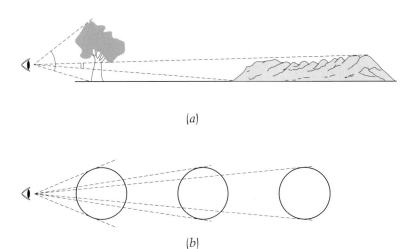

(a)

(b)

Figure 7.31: In (a) we see that a nearby tree can subtend a greater angle at the eye of the viewer than a much taller but distant mountain peak. In (b) we see the geometrical explanation of why the angle is small for a distant object. If a number of objects of the same size are placed at different distances, the angle subtended decreases with increasing distance as shown above.

identical light sources at varying distances in (a) a Euclidean universe and (b) an expanding non-Euclidean universe.

In the Euclidean universe, the apparent size steadily decreases for farther and farther sources. In the non-Euclidean geometry of the expanding universe, the result is rather unexpected. The angle subtended by the source first decreases and then *starts increasing with increasing distance.*

This means that, as shown in Figure 7.32(b), the sources start off by appearing smaller and smaller as their distances increase, but beyond a certain distance they start getting bigger! Thus the farther away a source, the bigger it would look. This decrease followed by increase would mean that at a certain distance the angle subtended by the source at the observer is a minimum. No source would look smaller than its size at this *minimum point.*

When Hoyle examined the different expanding models, he noticed that the distance of the minimum point is smaller the denser is the universe. Taking redshift as a measure of distance, for the Einstein–de Sitter model the minimum point lies at a redshift of 1.25, while for

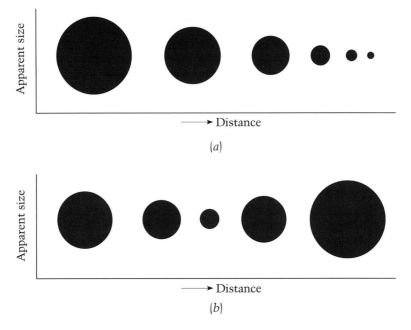

Figure 7.32: In (a) we see, in a Euclidean universe, a series of identical round light sources of varying distances from the observer. The farther away they are, the smaller they appear. By contrast, in (b) we find that a similar set of sources in an expanding universe may exhibit an anomalous behaviour. We see that first the sources get smaller and smaller as we look at more and more distant ones; however, beyond a certain distance, farther sources appear bigger! The point at which the source looks the smallest in size may be termed the minimum point.

type II models the minimum point lies at smaller redshifts. And for type III models the minimum point lies farther away, at redshifts larger than 1.25.

This apparent magnification of distant objects is simply another example of gravitational lensing, at which we looked in the previous chapter. Light rays coming from a distant source are bent by the matter in the universe lying en route in such a way that we see extended images of the distant sources.

Although this magnification test should, in principle, tell us what type of universe we live in, in reality it turns out not to be so definitive. The reason is that nature does not allow the astronomer the luxury of having a class of identical-sized sources. Whether galaxies, radio sources or quasars, the members of a population show enormous variation in intrinsic size. This makes it nearly impossible to spot the trend expected above and to locate the minimum point.

Nevertheless, the test is potentially so important that astronomers are continually tempted to search for a homogeneous population of sources to which it could be eventually applied with success.

RELICS OF THE BIG BANG

The idea of a vast universe extended on the scale of at least several tens of billions of light years and expanding is extraordinary enough. The implication that the entire structure arose out of a gigantic explosion requires further enormous leaps of the imagination. However, in a scientific approach to the problem one must dispassionately look for evidence that corroborates this picture.

A step in this direction was taken in the mid-1940s by the American physicist George Gamow. Gamow extrapolated the expanding universe models into the past and arrived at some very suggestive scenarios.

First, when he examined the then available evidence on the state of the universe, he found that at the present epoch it contains mostly matter and very little radiation. However, on mathematical extrapolation into the past, the relative importance of matter compared to radition drops. As we know, when a ball of gas is compressed, it gets more dense. So does a ball containing radiation: the density of radiation within it also grows. However, the density of radiation increases more rapidly than the density of matter. The calculations show that in the past, when the universe was ten times smaller than now, the density of matter was a thousand times higher than now. But, the density of radiation was *ten thousand times* what is is now. And this trend would continue as one extrapolated further into the past. So, Gamow argued that if we went far back into the past to when the universe was highly dense, we would find it dominated by radiation rather than by matter. And its temperature would correspondingly be very much higher than it is today.

Gamow's next extrapolation involved examining how a radiation-dominated universe would expand and how its temperature would drop with time in those early epochs. In particular, the era when the universe was passing through an age range of about one second to about three minutes interested him. For, during this era, the temperature of the universe dropped by a factor of about a hundred, from around ten billion degrees to around a few hundred million degrees. And at these temperatures, Gamow argued, the subnuclear particles, the neutrons and protons, could combine to form the nuclei of all the chemical elements that we see in the universe.

Recall that in Chapters 2 and 3 we encountered a similar range of temperatures in the cores of stars and saw how at these temperatures stars play the role of thermonuclear fusion reactors, generating energy *while making atomic nuclei*. Gamow hoped that similar developments would be possible in the early universe.

With his younger colleagues Ralph Alpher and Robert Herman (Figure 7.33 (a)–(c)), he went on to carry out the ambitious task of calculating how these fusion reactions would proceed in a rapidly expanding universe. In retrospect, and with the hindsight of further

Figure 7.33(a): George Gamow.

Figure 7.33(b):
Ralph Alpher.

Figure 7.33(c):
Robert Herman.

information about atomic nuclei, one can say that Gamow achieved partial success with his programme. We know today that the early universe can make light nuclei such as deuterium, tritium and helium, but not the heavier ones from carbon onwards. The process of primordial nucleosynthesis more or less stops at helium, the stable nucleus with two neutrons and two protons. Thereafter the process encounters unstable nuclei. The reason is similar to that discussed in the case of stars like the Sun, where there is the problem of crossing a barrier of unstable nuclei containing between five and eight particles. We saw that the problem was solved for stars by Fred Hoyle's solution of a resonant reaction that makes carbon from three nuclei of helium. Unfortunately that trick does not work in the early universe. Thus it became necessary to look to stars for making the nuclei from carbon upwards.

The microwave background

Nevertheless, another important prediction was made by Gamow, Alpher and Hermann. They predicted that in the aftermath of the process of primordial nucleosynthesis there will be a relic radiation background, which by today should have cooled off to a very low temperature. Not having any precise way to calculate this temperature, Alpher and Hermann reckoned that it would be around 5 K (five degrees on the absolute scale or −268 °C), while Gamow guessed a somewhat higher value of 7 K. But the important aspect of this radiation as predicted was that it should have a black body spectrum (see Chapter 2 for a description of the latter).

This prediction was more or less forgotten during the 1950s, when it was realized that primordial nucleosynthesis would not deliver all the chemical elements observed in the universe and that the bulk of them must be made in stars. So when in 1964 Arno Penzias and Robert Wilson accidentally detected an isotropic background radiation without any known source, at a wavelength of 7.3 cm, they were at a loss to understand its origin. Working at the Bell Telephone Labs in Holmdale, NJ, they had been, in fact, testing their horn-shaped antenna (see Figure 7.34) in order to measure radio intensities in the plane of the Milky Way. After deducting all possible contributions to the observed radiation this small but non-zero part remained. In order to trace whether its origin was due to some contamination, Penzias and Wilson even checked the antenna for pigeon droppings!

Figure 7.34: Penzias and Wilson with their horn-shaped antenna (courtesy of Bell Telephone Laboratories).

The news of their finding travelled to Princeton, where Bob Dicke and Jim Peebles were already planning studies of relic radiation. They had, however, arrived at their conclusions independently of the earlier work of Gamow, Alpher and Herman. They were able to recognize the relic radiation in the findings of Penzias and Wilson. Thus their paper, 'Measurement of excess antenna temperature at 4080 Mc/s', which appeared in the *Astrophysical Journal* in 1965, produced a sensation amongst cosmologists that was out of proportion to its rather mundane and modest title.

Penzias and Wilson had assigned a temperature 3.5 K to this excess radiation on the assumption that it was a black body type of radiation. The progress towards establishment of the full black body spectrum was slow but sure as several different groups stepped in to measure the radiation at different wavelengths. Figure 7.35 shows the most spectacular effort, that of the COBE (cosmic background explorer) satellite launched in 1989. A black body curve of temperature 2.7 K fits the data very well indeed. Because the bulk of the energy of this radiation lies in the microwave region, this radiation background is usually termed the *microwave background*.

Gamow's original idea of primordial nucleosynthesis also received a shot in the arm in the 1960s with the realization that the amount of helium in the universe, about a quarter of the entire observed mass, was far more than could have been produced by stars in the course of their evolution. Thus an extra source for this helium was needed and the

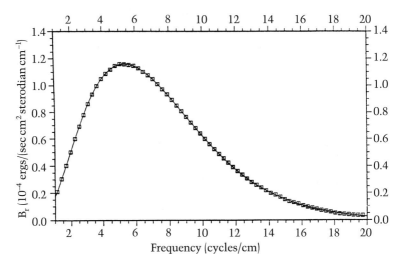

Figure 7.35: The black body curve fits very well the large number of measurements carried out by the Cosmic Background Explorer (COBE) satellite in 1989. This shows the cosmic background spectrum at the north galactic pole.

early universe provided just the right setting for it. The primordial production of helium could be as much as ninety per cent of what is observed, with the stars contributing the balance. A paper by Robert Wagoner, William Fowler and Fred Hoyle in 1967 gave a corrected and updated version of Gamow's earlier work and raised the credibility of the primordial nucleosynthesis scenario.

Astroparticle physics

This is why, since the 1970s, the idea of the big bang has enjoyed the status of 'most favoured theory' in cosmology. And cosmologists have become more daring in their extrapolations to even earlier epochs than those contemplated by Gamow. Somewhat ironically, the particle-theorist school, which had in the 1940s and 1950s considered Gamow's ideas too bizarre to be credible, has now joined the big bang bandwagon, in the so-called *astroparticle physics programme*. The basis of this programme is as follows.

Particle physicists are interested in looking for a unified theory of physics that encompasses all the known but different physical interactions, such as the electromagnetic interaction, the interactions that determine the behaviour of atomic nuclei and, of course, gravitation. Theoretical studies suggest that such a unification, if it really exists at all, will show up when material particles are interacting at extremely high energies.

Huge and powerful accelerators at CERN or Fermilab (Figure 7.36) are used by physicists to study the interactions of highly energized particles. However, the highest energies attained by these accelerators fall short of the target needed for unification by a whopping factor of over a million billion! In other words, the particle theorists have no hope of finding a laboratory that will ever test their unification theories. Unless . . .

Unless, they consider the expanding universe as their laboratory. For, as we examine the universe closer and closer to the epoch of the big bang, by looking at ever more distant galaxies, we find its temperature rising and, correspondingly, all the particles in it getting more and more

Figure 7.36: Particle accelerator at Fermilab in Illinois, USA.

energetic. Thus Gamow had found temperatures of the order of ten billion degrees a second after the big bang. A particle physicist would find a temperature a billion billion times higher at an earlier epoch when the universe was only a billionth of a billionth of a billionth of a billionth of a second old. Such an epoch will be of interest to the astroparticle theorist, because then the particles did have energies high enough for unification of all the important interactions except gravitation to be a reality. Thus, we may say that particle physicists also have a vested interest in big bang models.

Formation of large-scale structure

How does the cosmologist stand to gain from this collaboration? Well, the main problem that remains to be solved in cosmology is, how does the large-scale structure of the universe (see Figures 7.12.–7.16) come out of big bang models? Recall that in order to simplify all calculations, the universe was assumed to be *homogeneous*. Now we have to reexamine the possibility that initially it was not perfectly homogeneous, that initially it had tiny inhomogeneities, which grew into what we see as galaxies, clusters, superclusters, filaments and voids.

The particle physicist Sheldon Glashow has revived the Indian mythological concepts of a snake swallowing its tail (see Figure 7.2) by linking the largest and the smallest structures in the universe in the picture of a similar snake in Figure 7.37. Astroparticle physics will hopefully provide some credible starting conditions from which structures would evolve. The major research industry in cosmology today revolves around precisely this idea. And a major piece of evidence to latch on to amidst all these speculations is the discovery of tiny fluctuations in the microwave background radiation, of the kind first found by COBE.

For, the COBE satellite registered another spectacular success in 1992, when it was able to detect very fine-grained structure in the hitherto smooth-looking microwave background. This structure (see Figure 7.38) was in the form of ups and downs of the local temperature in the sky in different directions. On areas subtending an angle of approximately 10 degrees by 10 degrees, the COBE found that the temperature fluctuates by a few parts in a million!

Indeed, COBE's discovery initially brought a great relief to theoreticians, who were desperately searching for any sign of inhomogeneity in the relic radiation. For, it stands to reason that any fluctuations in

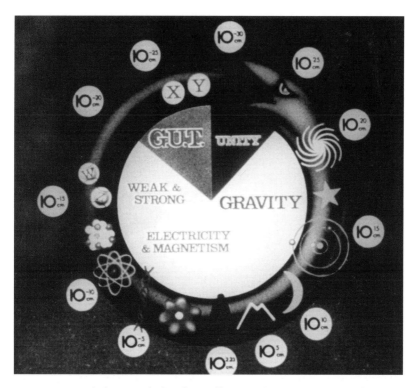

Figure 7.37: Glashow's snake has the smallest particle structures in its tail, getting progressively larger towards the mouth. The tail-swallowing suggests that the largest and the smallest structures are closely linked.

matter distribution (which grow into large-scale structures) should be coupled to like fluctuations in radiation. It would be hard to imagine that matter was inhomogeneously distributed but radiation was not. Prior to 1992, the earlier ground-based searches into the microwave background had failed to reveal any inhomogeneities.

The early euphoria of having found the ultimate evidence about the universe, however, has given way to more caution, as it is becoming clear that the problem of forming structures is still not all that easy. One important constraint is that one cannot afford to assume that *all* matter interacts with radiation. For that would have left greater imprints of matter-inhomogeneities on the radiation background, imprints far larger than found by COBE. So cosmologists have to invent a special form of matter which *does not* interact with radiation. Not

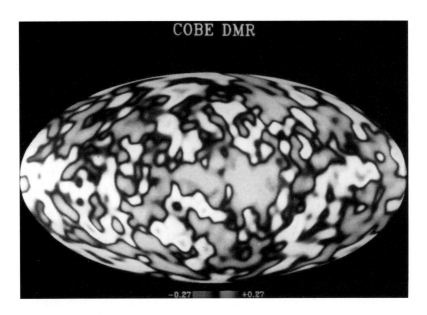

Figure 7.38: The pattern of inhomogeneity of temperature of the microwave background as revealed by COBE in 1992 (by courtesy of George Smoot, the COBE team and NASA).

only that, they have to assume that this strange kind of matter *makes up around ninety per cent of all matter in the universe*. Such matter would have no interaction with any kind of radiation and so would be dark to all kinds of light.

We now look at the evidence for dark matter, which has been accumulating from extragalactic astronomy. It will be interesting to see whether it is of the right kind and in the right amount as that required by the structure-formation scenarios.

DARK MATTER

Dark matter is a rather intriguing issue that may have very wide ranging implications for cosmology. As mentioned earlier, it is difficult to estimate the density of matter in the universe at the present epoch; however, should this become possible, we could tell whether we live in an open or a closed universe.

The problem esssentially is that astronomers are not sure whether the matter they *see* in the universe provides a good estimate of the total density. For there are definite indications that dark matter, not normally detectable through the various types of telescope, is present in the universe in substantial quantities.

The evidence comes in two different contexts, one in individual galaxies and the other in clusters of galaxies. Let us briefly look at it in that order. The key observation for galaxies comes from studies of the motion of clouds of neutral hydrogen. Such clouds move under the gravitational attraction of a galaxy just as planets move around the Sun.

Consider first a problem in our solar system. We know that the Earth goes round the Sun in one year. Can that information tell us the mass of the Sun? It can, provided we also know the Earth's distance from the Sun. Armed with this information, we can work out the Sun's mass, by using the law of gravitation. Next, if we do the same exercise for the orbit of Mars or Jupiter we will get the same answer.

This is hardly surprising, since the planets all move according to Newton's laws of motion and gravitation. As Kepler found before Newton, the planets move according to a set pattern, which he stated in his three laws of planetary motion. The Keplerian laws tell us that the speeds with which the planets go round the Sun progressively reduce as we look at farther and farther planets. The average speed of the Earth is a little over six times the average speed of Pluto, for example.

The Keplerian rules are also expected to apply to clouds of neutral hydrogen going round galaxies. We expect to find that a cloud farther away from the centre of a galaxy will move more slowly than a cloud comparatively nearer. Neutral hydrogen is known to radiate at 21 cm. Using this wavelength for observation, astronomers began to measure the average speeds of such clouds. To their surprise, they found results of the kind shown in Figure 7.39. *The speeds remain roughly constant over distances as far as three times the visible limit of a galaxy.* Why do they not get smaller and smaller for farther clouds?

Unless one renounces belief in the Newtonian laws and general relativity, one is led to assume that the gravitating matter that moves these clouds (as the Sun moves the planets around) extends well beyond the visible boundary of the galaxy. This is the dark matter in a galaxy. Its total mass is not negligible, it may even exceed the visible mass in the galaxy!

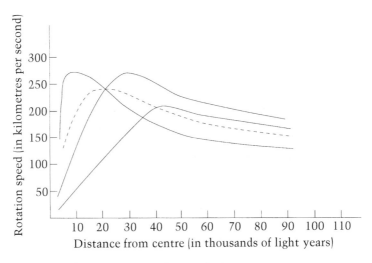

Figure 7.39: The rotation curves of some galaxies.

The second line of evidence comes from clusters of galaxies. Take the Coma cluster, shown in Figure 7.15. It has galaxies visible in the form of bright dots. One can measure their motion within the cluster and estimate how much energy resides in that motion. Again, if one believes that these galaxies have been moving under one another's attraction for a time long enough to have settled down to a state of dynamic equilibrium, then one can again estimate how much gravitating mass resides in the cluster. And the answer is that *as much as ten times the visible mass seen as galaxies is out there but it cannot be seen.*

Dark matter has posed problems to astronomers: they must make up their minds what this matter is made of. There are conventional options as well as non-conventional ones. For example, we could think of this matter in the form of planetary masses, like, say, the planet Jupiter. Or such objects could be bigger, but not massive enough to have become stars. A star like the Sun has sufficiently high temperatures at its centre to trigger off a fusion reaction there. But a ball of gas only a tenth of the mass of the Sun may not have a hot enough core. Such tiny objects will not be seen by normal observing means. They are generally called *brown dwarfs*. Or, one could think of dead stars, that is, neutron stars and white dwarfs, which have burnt out their nuclear fuel, or even black holes, which, of course, cannot be seen. All these objects are made of normal matter largely in the form of neutrons and protons. It is commonly called *baryonic matter* since neutrons and protons are broadly grouped as a class of particles called *baryons*.

These conventional options, however, are not very palatable to the big bang cosmologists. Without going into technicalities, one can say that such baryonic matter cannot exist beyond a rather low level in the universe. (Its density cannot exceed a few per cent of the closure density.) For, if it did, one would have problems in explaining the deuterium abundance observed in the universe. The process of primordial nucleosynthesis fails in producing adequate deuterium if the above density limit for baryonic matter is exceeded. Another problem is that baryonic matter interacts with radiation and if it formed more than a small proportion of the dark matter one could not then understand how galaxies and clusters could have arisen in the universe without disturbing the observed very smooth microwave background.

Both these issues are too technical to be elaborated here. Suffice it to say that big bang cosmologists take them seriously enough to rack their brains for alternative candidates for dark matter. Several esoteric options of *non-baryonic matter* have been suggested, such as massive neutrinos, photinos, gravitinos, axions, etc. These are particles conjectured to exist by those who are studying the ultimate microscopic structure of matter. Sometimes such particles are referred to as WIMPs (weakly interacting massive particles). So far none of them has been found in high-energy particle accelerators.

However, let us conclude this account of dark matter by mentioning one interesting way of looking for large-planetary-mass objects, brown dwarfs, dead stars etc., all falling within the normal baryonic option. This is the method of *gravitational microlensing*.

Figure 7.40 describes a typical microlensing event. Suppose we are observing a star moving across the halo of our Galaxy. If, in this process, a dark object comes close to its line of sight, the star may be gravitationally lensed by the dark object. This event will temporarily brighten the star as in Figure 7.41. By carefully monitoring these stars one may detect such gravitational dark objects.

Two experiments acronymed by MACHO (massive compact halo object) and EROS (expérience de recherche d'objets sombres) have been in operation and have registered successes in finding such objects. The question is, in what numbers are they present? Can all dark matter

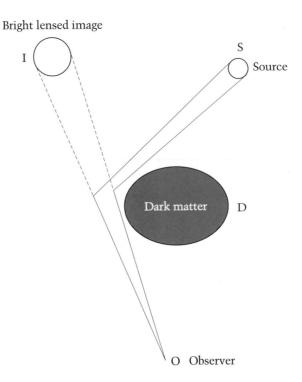

Figure 7.40: The geometry of a typical microlensing event.

be in this normal form as looked for here? Or do we need esoteric matter in abundance, as required by many scenarios of structure formation?

The future holds the key to these questions.

CONCLUSION

We have come a long way from the innocuous-sounding question, why is the sky dark at night?

From the darkness of the night sky to the search for dark matter is the long tale of modern cosmology, which today depends as much on probes of the most remote parts of the universe with the latest technological tools as on daring extrapolations of known science into the unknown

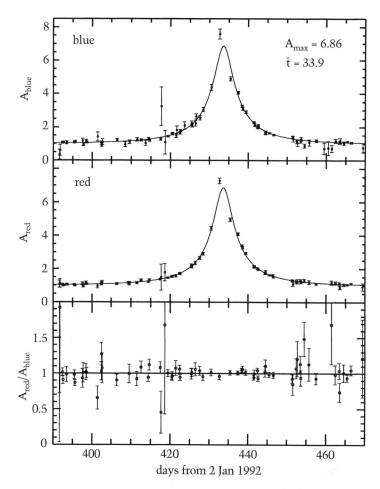

Figure 7.41: A characteristic rise and fall in the intensity of a star as it is microlensed by a dark mass (courtesy of the MACHO project).

past that get as close to the presumed epoch of creation as science would permit.

But, as J.B.S. Haldane once warned, *the universe is not only queerer than we suppose, it is queerer than we can suppose.* Who knows, but a surprise may be in store for us that will force us to change our present notion of the origin of the universe. After all, if the history of astronomy is any guide, we are just about due for a surprise.

Epilogue

In this book we have had a look at seven wonders of the cosmos. Once we leave the narrow confines of this planet, the vast universe around us presents us with vistas of increasing grandeur. That we are able to rationalize what we observe is itself a wonder.

Indeed, why should the science that we have picked up by experimenting on such a tiny location as the Earth, over no more than three centuries, successfully apply to phenomena spanning billions of light years of space and billions of years of time?

One could go further and raise a more philosophical question: Why should there be any laws of science at all, regulating the operations of the universe?

I will not enter into discussions of these questions. My purpose in raising them was not just to underscore the great success of human efforts against the backdrop of the immensity and complexity of the cosmos, but also to express caution that the science we so far know need not be complete. The cosmos may spring a few more surprises yet that will call for addition to our understanding of science itself.

We should not therefore be surprised if we find that there are still some puzzling features of the cosmos left to be sorted out. Indeed we should be disappointed if there were none.

Here I will list a few mysteries that still call for further thought and may even add to our understanding of basic science.

THE SOLAR NEUTRINO PUZZLE

The Sun is the nearest star and the one which we are able to observe and study much more closely than any other star. Yet there is a puzzle connected with the Sun that has so far defied solution.

As we discussed in Chapter 2, the Sun is currently generating energy through a chain of nuclear reactions which produce a large flux of neutrinos. Neutrinos can escape from the deep interior of the Sun easily, since they hardly take any notice of the matter around them. (Contrast this behaviour with that of photons, which get very much knocked about before they eventually emerge from the Sun.)

Figure E.1 illustrates the experiment set up deep underground in the Homestake mine by R. Davis to detect neutrinos coming from the Sun. Although this experiment has been in operation since around 1970, the result so far has been disappointing for theoreticians. The detector is not picking up as many neutrinos from the Sun as the theory of nuclear fusion would want us to believe. The discrepancy, with only about one-third of the predicted neutrinos being detected, is serious enough to cause worry. Is the detector not functioning properly? Is the theory of Sun's internal structure not quite correct? Is there a gap in our understanding of nuclear reactions? Or is our understanding of neutrinos still imperfect?

In the last two-and-a-half decades all these alternatives have been investigated but as yet no satisfactory explanation has emerged. In the meantime, new generations of solar neutrino experiments have started operating. Among these is one at Kamiokande in Japan, which looks for neutrinos scattered by electrons. The Kamiokande II experiment has 680 tonnes of ultra-pure water, which acts as the detector. The results from this experiment yield about half the expected number of neutrinos from the Sun. A more sensitive experiment called Super-Kamiokande is expected to provide further data.

Two detectors using gallium are known by the acronyms SAGE and GALLEX. They started producing results from 1991–2. Here, too, the observed flux of solar neutrinos falls far short of the expected value, being in the range 40–60 per cent.

The neutrinos being looked for in the different detectors fall in different energy ranges. All the experiments have some statistical uncertainties associated with experimental errors. However, even allowing for these, the discrepancy is serious.

So the ball is thrown back to theorists, particularly particle physicists, who are still trying to produce a unified scheme which will fit neutrinos of different species. It is possible that we may be able to

Figure E.1: The neutrino detector experiment pioneered by R. Davis deep underground consists of a huge tank of fluid called perchlorethylene (C_2Cl_4), which is exposed to the neutrinos coming from the Sun. The neutrinos interact with the chlorine nucleus in the solution and change it to argon, which can be detected. Thus by measuring argon nuclei we estimate the neutrino flux (photograph by courtesy of R. Davis Jr., Brookhaven National Laboratory).

understand and explain the above discrepancy after we have properly understood what neutrinos are.

In the 1980s, another useful probe of the solar interior came from the field of *helioseismology*. The subject arose from close studies of the disturbances of the solar surface. In fact, as early as the 1960s periodic disturbances were noticed with a period of five minutes in patches covering half the solar surface. Known as 'five-minute oscillations', these turned out to be the tip of the iceberg! The Sun also has seismic oscillations with much longer periods (20–60 minutes, 160 minutes etc.).

The oscillations arise from internal fluctuations of the Sun. Starting with a solar model one can deduce what type of oscillation one might expect to see, and then compare with what is seen. In this way we can check, change or confirm our postulates about the solar interior. Thus one finds that the external spin in the Sun's surface (which is like the Earth's spin about its polar axis) continues inwards but does not increase as rapidly with depth as some scientists had expected. Also, the abundance of helium inside is significantly high, higher than is convenient for theorists wishing to solve the solar neutrino problem just described.

Because of these dividends concerning our understanding of the Sun's interior, helioseismology has become an important field of study in solar physics.

THE FORMATION OF STARS AND PLANETS

Although we have described how stars form and how planetary systems develop about them, the scenario could be painted only with a broad brush. Thus several issues remain to be sorted out.

For example, what is the role of the magnetic field in this picture, keeping in view that stars and planets do have magnetic fields? Hannes Alfven and Fred Hoyle have given a plausible account of how the magnetic field could have played a role in transferring angular momentum outward from the sun to the planets. The magnetic lines of force linking the central part of the contracting cloud to the outer parts of the proto-planetray disc tend to slow the former down and make the latter spin faster (see Figure E.2). This is why the Sun itself spins rather slowly but

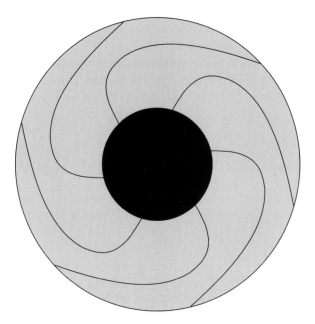

Figure E.2: The magnetic lines of force linking the central part of a collapsing cloud (which becomes the star) to the outer parts of the protoplanetary disc. These lines get wound up as the central part tries to spin faster. This generates a torque which pulls back and slows down the central part, at the same time making the disc spin faster.

the planets have a much greater speed of rotation round the Sun than would otherwise be the case.

There are still questions. Not all the planets spin in the same sense as they go round the Sun. The spin of Venus is almost opposite while the spin axis of Uranus is nearly perpendicular to its direction of rotation round the Sun. Why? And what does the asteroidal belt between Mars and Jupiter indicate? Is this the debris of a broken planet or bits and pieces that could not get together to form one? Both ideas have been proposed.

There are deeper questions about planets around pulsars! How did they form? Recall that pulsars are a late stage in a star's life, and the scenario of planets forming along with a new star does not apply here.

Finally, how common are planetary systems? This question has special significance for the search for extraterrestrial intelligence, as we will discuss at the end.

THE ENERGY OF QUASARS, RADIO GALAXIES AND ACTIVE GALACTIC NUCLEI

Until the 1920s, the secret of stellar energy had not been unravelled. Today we can say the same for the vast energy coming out of quasars, active nuclei of galaxies and radio galaxies (see Chapter 5). It is generally believed that in each of these cases the energy reservoir lies in the strong gravitational influence of a very compact object, ideally a black hole. This explanation has gained considerable popularity, but there are sceptics who question its efficacy. Let us hear their (minority) viewpoint.

To begin with, the dynamics of the surroundings of the central region show no evidence for the infall of matter, as would be required if a black hole were present there. Rather, the evidence points to the ejection of matter.

If a black hole were operating in the central region, its mass would decide its Schwarzschild radius. For a black hole of a billion solar masses the Schwarzschild radius would be three billion kilometres. An accretion disc around it will have a radius about a thousand times larger, that is, about one-third of a light year. To be able to *see* such a disc for a source located at a distance of, say, thirty million light years, the telescope needs to have a resolution of about one-thousandth part of a second of arc. This degree of resolution is far beyond the capacity of the best optical telescopes, including the HST. Thus the claims made of seeing the accretion disc do not really refer to the black hole accretion disc but a much larger disc or ring that may surround the central object in the same way as the protoplanetary disc surrounds a star. Thus the evidence for a central black hole is highly speculative, being based on the correctness of a string of assumptions.

The calculations of how energy is derived from the black hole and converted into radiation assume the best possible efficiency of whatever process is involved. For example, the gravitational energy of the black hole needs to be extracted and used to energize particles which spew outwards in a highly collimated jet. Then the energy of motion of these particles is to be converted to radio waves and other forms of radiation. It is not clear that high efficiency will be possible in these processes, since the same is not seen anywhere else in astronomy. Allowing for lower efficiency jacks up the black hole mass and makes the model less plausible.

For objects with a rapid variation of energy output, the size has to be small. This requirement conflicts with the above need for a larger black hole to allow for lower efficiency.

Indeed, the *prima facie* evidence points to the *ejection* of matter from a compact region which may or may not house a black hole. A possibility involving 'new' physics is that matter is being created and ejected here.

It is possible to give a mathematical description of how this can happen *without violating the law of conservation of matter and energy*. The trick is to allow for a new basic interaction, with negative energy and negative stresses, to operate in the region. As we found in Chapter 2, gravity itself has the character of a negative-energy interaction. Such an interaction leads to an explosive creation and ejection of matter from the compact region.

We will shortly connect up this idea of a *minibang* with cosmology and the expanding universe.

THE REDSHIFT PUZZLE

Throughout this book we have assumed that the redshift of any galaxy or quasar or, for that matter, of any object lying outside our Galaxy is due to the expansion of the universe alone. Of course, superimposed on the expansion there may be small Doppler redshifts arising from random motions of galaxies or quasars in clusters. But these are expected to be small and the bulk of the redshift will be of cosmological origin. We will term this statement as the *cosmological hypothesis* (CH in brief).

Over the last three decades, however, some evidence has been gathered to suggest that there may be something wrong with this hypothesis. Although the CH seems securely based so far as galaxies are concerned, some astronomers have called into question its applicability to quasars.

Suspicion as to the cosmological origin of the quasar redshift began in the early 1970s, first from the realization that, unlike the case of galaxies, for quasars there is no clear relationship implying that quasars with larger redshifts are fainter. There is far too much scatter in the data to reveal any Hubble-type relationship. Indeed, it is hard to imagine that Hubble would have been able to arrive at a velocity–distance relation if he had found quasars first. See Figure E.3 for the Hubble diagram for quasars.

Since the mid-1970s, Chip Arp, himself a student of Hubble and a distinguished astronomer in his own right, has been finding evidence that does not fit in with Hubble's law. We give examples of three different types of evidence.[16]

Figure E.4 shows three quasars near a galaxy. Are they physically close to the galaxy, or are they in fact distant and happen to be randomly

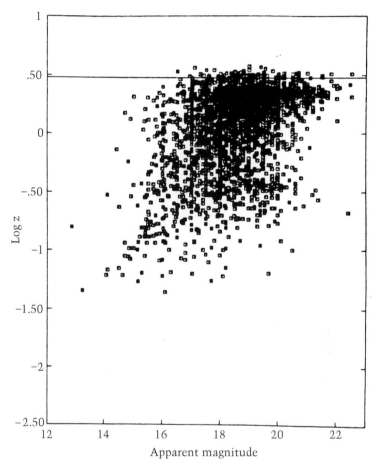

Figure E.3: The Hubble diagram of some 7000 quasars shows a scatter rather than a Hubble-type redshift–distance relation (courtesy of A. Hewitt and G. Burbidge). Vertical axis, logarithm of the redshift; horizontal axis, distance.

[16]To those interested in the details, we recommend Arp's very readable account of such cases in his book *Quasars, Redshifts and Controversies*, Berkeley, Interstellar Media (1987).

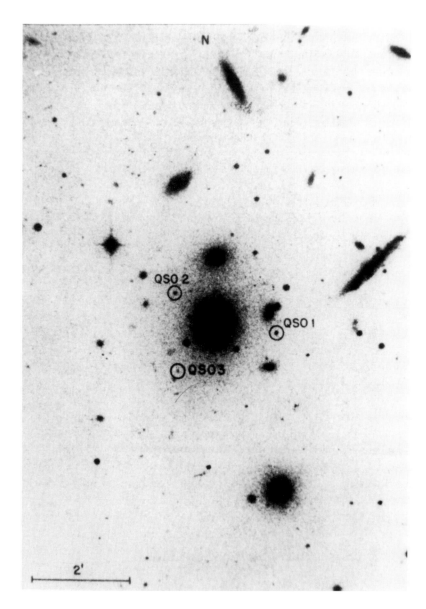

Figure E.4: Three quasars near the galaxy NGC 3842. Their angular separations from the galaxy are 73, 59 and 73 seconds of arc. Do they appear so close to the galaxy by chance? (courtesy of H. C. Arp).

projected in directions close to that of the galaxy? We will call these interpretations options (1) and (2).

Quasars are relatively rare objects and as such the sky is sparsely populated with them. We can estimate the probability that these three quasars happen to be projected close to the galaxy by chance. It works out at less than one in a million. Put differently, the event is as rare as getting a run of 20 heads *in succession* when we toss a coin. At such a low probability the statistician would have to conclude that the second option is unlikely and that the quasars are physically associated with the galaxy.

But option (1) goes against Hubble's law, which states that redshift depends only on distance. Here we have a galaxy with redshift as low as 0.02 physically associated with quasars of redshifts 0.34, 0.95 and 2.20. Does this mean that most of the redshift of each quasar is not due to the expansion of the universe? Does it have an extra, intrinsic, component?

Figure E.5 shows two triplets of quasars with different redshifts but each triplet well aligned. Both the triplets were found on the same photographic plate. The probability of this happening by chance is as low as getting a run of 12 heads in succession by tossing a coin (the odds against this being about 4000 to 1). The odds against such a system coming about by chance will be longer still if we notice the matching of redshifts of the three quasars in the two triplets: the two sets of end redshifts match, as do the central redshifts. Normally such well-aligned systems are expected to be physically associated. Recall that we have radio lobes aligned across a central galaxy in a radio source. By analogy we may think of these quasars to be so aligned because of an ejection process. For example, the central quasar in each triplet may have ejected the other two in opposite directions. The redshifts may then be due to the Doppler effect of local ejection.

Finally, in Figure E.6 we see a case where two galaxies of different redshift seem to be connected by a thick filament. The large galaxy has a redshift 0.029 while the small one, in the lower left of the photograph, has a redshift 0.057. If we assume that the connection is real, then the smaller galaxy has a relative radial velocity of about 8300 km per second. This is too high to be explained as a random relative motion. So, in order to keep Hubble's law alive in this case, we have to assume that the two galaxies *are not connected*, that the smaller galaxy happens to be projected just at the end of the filament issuing from the big galaxy!

In all these cases, we have to invoke rather contrived projections to explain what we see. There have been attempts to explain these asso-

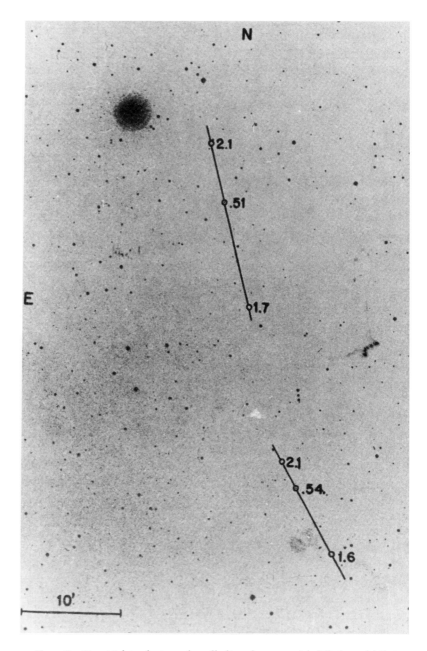

Figure E.5: Two triplets of extremely well-aligned quasars with differing redshifts as marked. If Hubble's law is correct, these alignments must all be due to pure chance (courtesy of H. C. Arp).

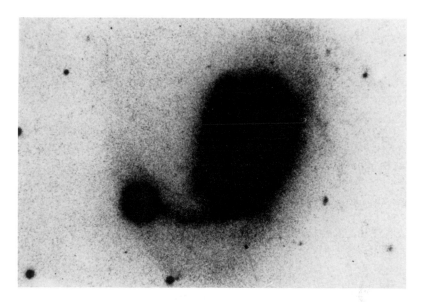

Figure E.6: The large galaxy NGC 7603 with a companion galaxy (lower left) apparently attached to it by a thick filament. The two have different redshifts, the difference being too large to be explicable by random motions (courtesy of H. C. Arp).

ciations of discrepant redshift objects as examples of gravitational lensing, but these explanations too do not carry much conviction.

We may also recall from Chapter 6 the explanations offered for the apparent superluminal motions in quasars. We then pointed out that there is no superluminal motion if the quasars are actually closer than required by Hubble's law.

Are such cases, therefore, examples of very rare occurrences, as Hubble's law would have us believe, or are we seeing in them a hint that some new physical idea is needed to understand these anomalies? Again, let me emphasize, the majority attitude towards such anomalous phenomena has been to ignore them, rather than probe them further.

WAS THERE A BIG BANG?

The notion of the big bang arose by extrapolating the observed expansion of the universe far back into the past, using Einstein's general theory of relativity. As evidence of the primordial event, one cites the microwave background observed today and the abundances of some light atomic nuclei, which cannot be explained in terms of nucleosynthesis in stars. But, despite these positive points, the big bang notion may not in fact turn out to be right. There are several reasons for holding such an agnostic view.

Firstly, the concept of the big bang itself, that is, an epoch of spacetime singularity, is one that defies any physical study. At this epoch, the density and temperature of matter and radiation become infinite, all volumes shrink to zero and the geometrical properties of spacetime become undefined. The big bang has thus acquired a mystical aura which is totally out of place in a scientific theory. Normally, if a physical theory leads to unwanted infinities or zeroes for physical quantities, it is suspect and attempts are made to improve it so that the unwanted elements are eliminated. It is therefore essential to have a modified version of Einstein's theory which will get round the singularity problem. If a successful merger between quantum theory and general relativity is made, this theory may well be able to do away with the big bang problem.

Assuming that we keep the physics confined to the post-big-bang era, the age of the universe can be calculated. It turns out to be in the range 8–12 billion years for the Einstein–de Sitter model, taking into consideration the present uncertainty about the true value of Hubble's constant. There is a serious conflict, however, between this value and the ages estimated for some of the oldest stars in the Galaxy, which are in the range 13–17 billion years. How can the universe be younger than its contents? The ages of type II models (closed universe) are even shorter. Currently, attempts are being made to resolve the difficulty by invoking type III models which are of low density. However, the problem is that there are further constraints from other observations.

These other observations include the abundance of deuterium (an isotope of hydrogen commonly known as 'heavy hydrogen') in the universe, the fluctuations found in the microwave background by the COBE satellite and other detectors, the actual observations of the large-scale structure (galaxies, clusters, superclusters and voids), the occurrence of fully formed galaxies at high redshifts and the abundance of rich (i.e., more densely populated) clusters. It would take us into too many technical details to describe and assess these constraints. We can say, though, that in general experts now agree that these constraints have made it necessary to introduce additional parameters into the big bang picture. One such parameter is the *cosmological constant*.

This constant was introduced into general relativity in 1917 by Einstein, as he needed a cosmic force to balance gravitation, in order to produce a static model of the universe (see Chapter 7). The cosmological constant essentially specified the magnitude of this repulsive force between any two galaxies separated by a given distance. Later, he abandoned it as unnecessary, once it became established that the universe is not static but expanding. This constant is currently being resurrected in order to prop up the big bang scenario.

Rather than carry out a patchwork exercise like this, perhaps the time has come to reassess the evidence and think of an altogether different approach. One such idea currently under discussion is the *quasi-steady-state cosmology* (QSSC), proposed in 1993 by Fred Hoyle, Geoffrey Burbidge and the present author.

In this cosmology, matter creation in the universe is not relegated to a mystical event like the big bang, but is part of a well-defined field theory. Again sidestepping the technical aspects, one may say that in the QSSC the expansion of the universe is driven by a distribution of local creation centres or *minibangs*. These are around highly compact massive objects, which are close to the black hole state but are not actually black holes. Creation of matter is facilitated in the strong gravity of these objects. Not only that, the field acting as the agency for the creation of matter ejects this matter with great force, leading to an explosive situation. Phenomena such as quasars, active nuclei of galaxies and radio sources may very well be energized by this process.

The impact of these processes on cosmology is to make the universe expand. However, the creation activity may not hold steady; it may have ups and downs, leading to fluctuations in the steady expansion of the universe. As shown in Figure E.7, the expansion is quasi-steady, with alternating periods of expansion and contraction, much like ups and downs in a growing economy. The universe itself has no beginning, no end, no big bang, no big crunch: it goes on for ever. It is free from the problem of singularity.

This cosmology naturally does not have the problem of age. Very old and very young stars can exist in it without causing any embarrassment!

The particles created in the minibangs are the so-called *Planck particles* whose mass is determined by the three fundamental constants, viz. the speed of light, the Planck constant and the Newtonian constant

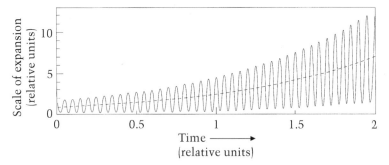

Figure E.7: Figure shows the variation with time of the scale factor for the QSSC. The dotted curve shows the steady expansion on which oscillations (continuous line) are superposed.

of gravitation. Light nuclei are produced as the decay products of the Planck particle, and their abundances, as calculated by the QSSC, match the observed findings very well.

The microwave background is explained as the relic light left behind by the stars of previous cycles. The model correctly predicts the present temperature of the background.

How can we distinguish between this cosmology and the big bang cosmology?

There are a few crucial tests. For example, if we are looking at some sources in the past epoch of the QSSC when the previous oscillatory cycle was close to its maximum extension, then these sources should be *blueshifted*. That is, the lines in their spectra should appear with increased frequencies compared to the laboratory values. Finding such cases is not easy because the sources in question will be very far away and faint. However, evidence of this kind would be impossible to explain in the standard big bang cosmology.

Other evidence to look for is in the form of low-mass stars which have just become red giants. Stars of, say, half a solar mass will burn very slowly and take about forty to fifty billion years to become giants. In the QSSC we should find such stars, born in the previous cycle. But in the big bang cosmology they cannot be accommodated.

The QSSC does not require dark matter to be made of esoteric particles like WIMPs. If it turns out that dark matter is largely made up of normal (i.e., baryonic) particles then this will be a point against the big bang.

FORMATION OF LARGE-SCALE STRUCTURE

Whatever cosmological model we choose, it will have to explain the existence and hierarchy of the large-scale structure that we outlined in Chapter 7. Indeed, in big bang cosmology, the above question is currently occupying centre stage amongst theoreticians. Did the structure form in an upward hierarchical sequence (the bottom-up scenario), with galaxies forming first, then getting into clusters which later formed super-clusters? Or did the reverse happen (the top-down scenario)? How much dark non-baryonic matter is needed to form the observed structures? Is it 'hot' or 'cold'?[17] Did (or does) the distribution of dark matter mimic that of visible matter? Did the cosmological constant play a role in the entire scenario?

All these variations are being tried, on the central theme that the gravitational force plays the key role in forming structures. The proof of the structure pudding lies in matching with the data observed today, both for galaxies and for radiation. That no successful contender has emerged is an indication of the complexity of the problem.

Or it may be an indication that the central theme is inadequate or wrong. The proponents of the QSSC, for example, would place their eggs in the minibang basket, arguing that structures develop through explosive matter creation around compact massive objects.

Future observational studies of the microwave background and large-scale structure may help settle the issue by providing additional detail. They may also make life more difficult for the theoreticians, of course!

THE SEARCH FOR EXTRATERRESTRIAL INTELLIGENCE (SETI)

I have kept one very interesting issue out of the scope of the seven wonders. The question 'Are we alone in the universe?' is being asked by lay persons as well as scientists. Indeed, with our Galaxy containing about a hundred billion stars such as the Sun, many of which might have planets, the question does assume significance. Our account of the cosmos would be incomplete without mention of the search for extra-terrestrial intelligence (SETI) project.

[17]Hot dark matter and cold dark matter are technical terms distinguishing the speeds of dark matter particles in the early stages of structure formation. The HDM particles were moving fast, the CDM particles were slow.

In Chapter 3 we referred to molecular clouds. The astronomy of millimetre-long waves has been successful in identifying complex organic molecules, the likes of which are found in life forms on the Earth (including us humans). Thus the tantalizing possibility exists of life forms elsewhere, built out of these jig-saw pieces. The expectation is that, since life requires energy and a congenial environment, it may develop on a suitable planet going round a star, which acts as the energy source.

There are sceptics, of course. For example, we still do not know how life came about on the Earth. What are the chances of its developing elsewhere? Are the chances sufficiently large to more or less guarantee the existence of another living system elsewhere? The sceptics think not, and would like to believe that we are indeed alone.

Then there are those who would prefer the empirical approach of search to theoretical arguments about whether extraterrestrial life is possible. Our present technology is just about capable of catching radio signals, if such are being exchanged between advanced extraterrestrials. The most likely wavelength for such interstellar transmission is believed to be the 21-centimetre waveband around the neutral hydrogen wavelength (see Chapter 7). Not only is it all-pervading in the Galaxy and should be familiar to our advanced neighbours (who will have worked out the physics behind it), it is also the case that the transmission on this wavelength does not suffer from much attenuation.

So, the empirical SETI approach is to open out our antennas and hope for some intelligent signals to arrive. Then, if we are successful in interception, we could initiate our own dialogue with the extraterrestrials.

And, if they are really advanced, they might be able to set us right on the puzzles we are discussing here.

CONCLUSION

Whatever extraterrestrials tell us, or we discover ourselves, the cartoon shown in Figure E.8 has the last word on how astronomy has advanced through the unexpected. The human prejudice that whatever is known at the current epoch should be sufficient to understand all the mysteries of the universe resists new ideas. Despite this resistance, they do break in, in an unexpected fashion. Therein lies the thrill in working in this

Figure E.8: Even if you
lock the door in many
different ways, an
intruder can still get in
by an unexpected
route!

field. The wonders that are unexpected are more wonderful than the
wonders that are expected.

So let us not speculate what the Eighth Wonder will be . . .

Index

in binary, 139–44
pulsar, 128 (*continued*)
  model of ,131–3
  planets around, 153–4
puranas, 248
Pythagoras theorem, 157, 164
Pythagoreans, 250, 251

quantum degeneracy, 83
  degeneracy pressures, 83, 85, 105, 183
quantum of energy, 51
quantum theory, 41, 47–62, 312
quantum tunnelling, 52–3
quantum uncertainty: *see* uncertainty
    principle
quasar (quasi-stellar radio source),
    198–203, 227
  alignments of, 309, 310
  as ejection phenomena, 313
  association with galaxies, 308, 310
  lensing of, 233–40
  VLBI measurements of, 209–12, 244–5
quasi-steady-state cosmology (QSSC),
    313–15
quasi-stellar object (QSO): *see* quasar

radiation damping of a charge, 151
radio sources, 195–7
Rahu, 29
rainbow, 14, 15
Readhead, A.C.S., 210
red giants: *see* giant stars
red lights for hazard warning, 15–16
redshift, 188, 212, 233, 268
  anomalous, 306–11
redshift–faintness relation, 275–6
  for quasars, 306–7
Rees, Martin, 200, 214, 216
refraction, 13–14, 24, 225
Refsdal, S., 225
Reifenstein, E.C., 137
resonance, 79–80
resonant reaction, 79, 80, 102
Revati, 155
Riemann, 173
Rigveda, 247
Ring Nebula, 43, 104
Roberts, D., 234
Robertson, H.P., 278
Roche, E., 141
Roche lobe, 141
Russell, Henry Norris, 38, 39

Saha, Meghnad, 55, 56
Saha's formula, 56–7
Sandage, Allan, 275
Sanduleak, 112, 113
Sarton, George, 93
scattering of light, 13, 14
Schmidt, Maarten, 198
Schwarzschild, Karl, 179, 180, 218, 219
Schwarzschild barrier, 189, 190, 191
Schwarzschild radius, 190
Schwarzschild solution, 179, 218, 236
scintillation, 124
search for extraterrestrial intelligence
    (SETI), 304, 315–16
Shaffer, D.B., 211, 213
Shapley, Harlow, 270, 273
Shelton, Ian, 111, 112, 113
Shesha Naga, 249
Shimmins, A.J., 198
shock wave, 106, 120–3
singularity, 189, 190, 191, 312, 313
Slipher, V.M., 267
Small Magellanic Cloud, 112
Solar Max satellite, 113
solar neutrinos, 300–3
spacetime, 166
spacetime diagram, 164
spacetime singularity: *see* singularity
special relativity, 85, 160–70, 206, 214
  contraction in space, 160–1
  dilatation of time, 162, 163
  conflict with Newtonian gravitation,
      170–3
spinar, 200
Spitzer, Lyman, 195
Staelin, D.H., 137
stars
  as black bodies, 62
  as energy source for life, 316
  energy of, 64–78
  exploding: *see* supernova
  formation of, 114–18
  luminosity of, 38, 40, 42
  physical characteristics of, 40–62
  pressure inside, 69–72
  sizes of, 63–4
  spectrum of, 47–51, 56
  surface temperature of, 38, 40, 57
  trajectories of, 4–5
  twinkling of, 24, 25
Steinberg, D.R., 142, 154